Management of Lipids in Clinical Practice

Third Edition

Neil J. Stone, MD
Professor of Clinical Medicine — Cardiology
Northwestern University Medical School
Chicago, Illinois

Conrad B. Blum, MD
Professor of Clinical Medicine
Columbia University College of
Physicians and Surgeons
New York, New York

Edward Winslow, MD
Associate Professor of
Clinical Medicine — Cardiology
Northwestern University Medical School
Chicago, Illinois

Published by
Professional Communications, Inc.

For orders, please call
1-800-337-9838

ISBN: 1-884735-59-2

Printed in the United States of America

DISCLAIMER

The opinions expressed in this publication reflect those of the authors. However, the authors make no warranty regarding the contents of the publication. The protocols described herein are general and may not apply to a specific patient. Any product mentioned in this publication should be taken in accordance with the prescribing information provided by the manufacturer.

This text is printed on recycled paper.

DEDICATION

To Karla, and to Scott, Adam,
Lauren, and my parents.

NJS

To Cindy, and to Julia and Harold,
and my parents.

CBB

To my teachers, family, and students
who have also taught me.

EW

ACKNOWLEDGMENT

The authors gratefully acknowledge the editorial expertise of Phyllis Jones Freeny, the graphic design work of Nikki D. Weaver, and the advice of Malcolm Beasley. NJS also wishes to acknowledge with gratitude the support of the Jacques M. Smith Distinguished Physician Award from Northwestern Memorial Hospital.

Table of Contents

TABLES

FIGURES

1 Introduction

Overview

Why would a book on managing lipids in clinical practice be written? It has been more than 12 years since the National Heart, Lung, and Blood Institute launched the National Cholesterol Education Program (NCEP). The announced goals of that program were to:

- Reduce the prevalence of high blood cholesterol in the United States
- Contribute to the reduction of mortality from coronary heart disease (CHD).

Initially, there were guidelines from the Expert Panel on the Detection, Evaluation, and Treatment of High Blood Cholesterol in Adults (ATP)-I. The panel recommended screening guidelines employing blood cholesterol values which could be fasting or nonfasting. Results from the 6-year follow-up of the cohorts from the Multiple Risk Factor Intervention Trial defined the desirable range as under 200 mg/dL, borderline being 200 to 239 mg/dL, and high being 240 mg/dL or greater. Specific algorithms for evaluation and treatment were provided. Low-density lipoprotein cholesterol (LDL-c) was chosen as the primary target for intervention. High-risk patients were identified as those having LDL-c >160 mg/dL or >130 mg/dL with two or more risk factors. LDL-c values under 130 mg/dL were considered desirable. For the first time, physicians evaluating hypercholesterolemia could refer to specific goals to be attained by dietary and, if needed, drug therapy. Although high-density lipoprotein cholesterol was considered a major risk factor, it was not considered for screening purposes at that time, in great part due to concerns over imprecise clinical measurement and lack of a science base to support its widespread utilization. With the publication of the Population Panel Report, there was in place both a population approach designed generally to lower cholesterol in the population and a high-risk strategy designed to focus on those at highest risk for CHD.

This extensive effort had a considerable impact. Data collected from the third National Health and Nutrition Education Survey (NHANES III) showed that the proportion of adults with high blood cholesterol (>240 mg/dL) fell from 26% to 20% while the proportion with desirable levels of blood cholesterol (<200 mg/dL) rose from 44% to 49%. Because of scientific progress resulting from an explosion of basic as well as clinical research, the ATP II was published in 1993. This report refined and expanded the previous set of guidelines. Some of the notable features are set forth in Table 1.1; these will be expanded in a later section on diagnosis.

The goal, then, is to include in one convenient reference much of the information that the practicing physician needs in order to manage those patients with abnormal blood lipids and lipoproteins. Although certain aspects of screening, evaluation, and treatment remain controversial, scientific progress has identified effective treatment of high blood cholesterol in those with established CHD to be a priority for the practicing physician. As will be shown in later chapters of this book, this aspect of "secondary prevention" has the ability to reduce:

- Cardiovascular morbidity and mortality
- Need for revascularization procedures
- Total mortality.

In addition, the practicing physician needs more information about how to identify, evaluate, and treat

TABLE 1.1 — NEW FEATURES OF THE ADULT TREATMENT PANEL II GUIDELINES
• Emphasis on risk status as a guide to the type and intensity of lipid-lowering therapy – Low risk: Those with less than two risk factors – High risk: Those with two or more risk factors but no overt coronary heart disease (CHD) – Very high risk: Those with known atherosclerotic vascular disease • Emphasis on high-density lipoprotein cholesterol, a major risk factor for CHD, in screening • Emphasis on physical activity and weight reduction to enhance dietary therapy • Emphasis on high-risk postmenopausal women and high-risk elderly who are otherwise healthy as candidates for cholesterol-lowering therapy
Second Report of the National Cholesterol Education Program (NCEP) Expert Panel. *JAMA*. 1993;269:3015-3023.

susceptible patients who do not yet have CHD. The potential of "primary prevention" to reduce the toll from CHD in the future makes this a compelling subject worthy of the attention of the practicing physician. An important challenge, however, is to do this in a cost-effective manner without compromising the main objective of reducing CHD deaths.

This effort is a considerable challenge. Utilizing population data from 1990, Sempos and colleagues attempted to analyze the magnitude of the task before US physicians. They estimated that approximately 52 million Americans 20 years of age and older are candidates for dietary therapy. This statistic suggests that approximately 29% of adults are candidates for at least dietary therapy. While this is less than the 36% of adults estimated to require dietary therapy using the previous NHANES II data, the job is still a formidable one (Table 1.2).

TABLE 1.2 — PERCENTAGE OF US POPULATION AGED 20 YEARS AND OLDER WHO ARE CANDIDATES FOR DIETARY INTERVENTION

Population	LDL 130-159 mg/dL >2 Risk Factors (%)	High-Risk LDL >160 mg/dL (%)	CHD and LDL >100 mg/dL (%)	Total (%)
All persons	7	16	6	29
Black	7	15	5	27
White	7	17	6	30
Mexican-American	5	12	4	21
Age 65-74 years	14	24	15	53

Abbreviations: LDL, low-density lipoprotein; CHD, coronary heart disease.

Adapted from Sempos CT, et al. *JAMA*. 1993;269:3009-3014.

Moreover, assuming that dietary intervention can reduce LDL-c by 10%, as many as 12.7 million Americans may be candidates for lipid-lowering drug therapy. This estimate includes approximately 4 million adults with established CHD for whom control of LDL-c is a priority. Of this number, 2 million or more are estimated to be 65 years and older.

Treating the Patient With Hyperlipidemia

Physicians involved in evaluating and treating the patient with hyperlipidemia must first consider the following:

- Are the lab results believable? The physician should understand the factors that affect blood lipid values.
- If abnormal values are seen, are there secondary causes which might explain these values? Diet, drugs, diseases, and disorders of metabolism all present possibilities.
- Is there evidence of a familial lipid disorder? For those with severe abnormalities or associated premature CHD, family screening is mandatory.
- What is the absolute and near-term risk of CHD or, when confronted with a case of severe hypertriglyceridemia, the risk of acute pancreatitis in this patient? Those who are young without other risk factors for CHD often do not need to undergo the same workup and treatment given to patients with similar LDL-c values but with either CHD or strongly positive risk-factor profiles that include a family history of premature CHD or diabetes.

Defining risk status is an essential part of the clinical approach. In some cases, this will mean that specialized lipoprotein tests are ordered. In every case, it means that a careful history and critical evaluation of pertinent exam and laboratory data should precede treatment, which, by necessity, must be lifelong. In great part, successful compliance will depend on the patient's understanding the following:

- Why treatment is indicated
- What the goals of therapy will involve
- How the appropriate follow-up and feedback on condition will be handled.

Management of lipid disorders will involve a wide spectrum of health care workers. This includes those in primary care medicine, internists, and cardiologists, as well as specially trained nurses, dietitians, and exercise physiologists. If the goals of therapy are clearly defined for each patient, each member of the team can contribute toward helping the patient achieve these goals. The value of the physician's endorsement and encouragement cannot be overstated.

Despite reports of greater cholesterol awareness among physicians, there is still much work to be done. A recent study of adult patients receiving lipid-lowering therapy showed that only 38% of patients achieved LDL-c target levels per NCEP ATP II (Pearson, 2000). Paradoxically, success rates were lowest among those felt to benefit the most—the high-risk subjects. Thus, LDL-c goals were met in 68% of low-risk patients, 37% of high-risk patients, and 18% of CHD patients. Drug therapy was significantly more effective ($P \leq 0.001$) than nondrug therapy in all groups.

The high proportion of patients who do not reach LDL-c target levels spans all socioeconomic classes. Not surprisingly, those with the highest level of education were more likely to reach their LDL-c goal than the less educated. The success rate was low for minority groups, with African American subjects less likely to reach their LDL-c goals than the white or Hispanic subjects. This finding was supported by the results of the Heart and Estrogen/Progestin Replacement study in which nearly half of African American women failed to use lipid-lowering medication (Schrott, 1997). This was significantly higher than the results seen in white women in this study.

Thus the challenge is to identify, evaluate, and treat those at the highest risk. There are significant barriers that must be overcome to achieve this goal. If this book encourages and advises health care workers to greatly improve this situation, its goal will be realized.

SUGGESTED READINGS

National Cholesterol Education Program Expert Panel. Report of the National Cholesterol Education Program Expert Panel on Detection, Evaluation, and Treatment of High Blood Cholesterol in Adults. *Arch Intern Med.* 1988;148:36-69.

Pearson TA, Laurora I, Chu H, Kafonek S. The lipid treatment assessment project (L-TAP): a multicenter survey to evaluate the percentages of dyslipidemic patients receiving lipid-lowering therapy and achieving low-density lipoprotein cholesterol goals. *Arch Intern Med.* 2000;160:459-467.

Schrott HG, Bittner V, Vittinghoff E, Herrington DM, Hulley S. Adherence to National Cholesterol Education Program treatment goals in postmenopausal women with heart disease. The Heart and Estrogen/Progestin Replacement Study (HERS). The HERS Research Group. *JAMA*. 1997;277:1281-1286.

Sempos CT, Cleeman JI, Carroll MD, et al. Prevalence of high blood cholesterol among US adults. An update based on guidelines from the second report of the National Cholesterol Education Program Adult Treatment Panel. *JAMA*. 1993;269:3009-3014.

Summary of the second report of the National Cholesterol Education Program (NCEP) Expert Panel on Detection, Evaluation, and Treatment of High Blood Cholesterol in Adults (Adult Treatment Panel II). *JAMA*. 1993;269:3015-3023.

2 Pathophysiology of Hyperlipoproteinemias

2

Hyperlipoproteinemias are responsible for producing elevations in the plasma levels of cholesterol and triglyceride. In clinical practice, the most important of these conditions are those resulting in elevation of low-density lipoprotein (LDL) because these commonly occurring disorders markedly increase the risk of coronary heart disease (CHD). For this reason, the disorders resulting in elevations of LDL are described in greatest detail.

Normal Lipoprotein Metabolism

Plasma lipoprotein metabolism can go awry in a variety of ways to produce abnormalities in the plasma levels of the lipoproteins. In order to understand these abnormalities, knowledge of important principles in normal lipoprotein metabolism is required.

■ Plasma Lipoproteins and Apolipoproteins

The single most important concept for understanding the disorders of lipid transport is that the lipids of plasma circulate in lipoprotein particles. Because of their insolubility in aqueous solutions, the lipids cannot circulate alone. The plasma lipoproteins are spherical particles with a nonpolar core composed of triglyceride and cholesteryl ester. A polar outer coat is composed of unesterified cholesterol, phospholipid, and specific proteins termed apolipoproteins. The apolipoproteins stabilize and impart solubility to the lipoproteins, are involved in catalyzing or modulating intravascular changes in the lipoproteins, and facilitate lipoproteins' entry into and exit from cells.

The major lipoproteins are usually categorized in terms of their density (Table 2.1) into four major classes: chylomicrons, very low-density lipoprotein (VLDL), LDL, and high-density lipoprotein (HDL). Chylomicrons, the transport vehicle for dietary fat, are usually 90% to 95% triglyceride by weight, 1% to 2% cholesterol, and <1% protein. VLDL, the carrier of endogenously produced triglyceride, secreted by the liver, is approximately 60% triglyceride, 12% cholesterol, 10% protein, and 18% phospholipid. LDL, a catabolic product of VLDL and normally the major carrier of cholesterol in the plasma, is 50% cholesterol, 25% protein, 20% phospholipid, and less than 5% triglyceride. HDL is the densest of the major classes of lipoproteins because of its high protein content; it is approximately 50% protein, 20% cholesterol, and 25% phospholipid.

■ Genesis and Fate of Chylomicrons

Dietary triglyceride in the intestinal lumen undergoes hydrolysis catalyzed by pancreatic lipase in the presence of bile salts to form monoglyceride and unesterified fatty acids. These products of lipolysis, together with bile salts and other lipids (eg, cholesterol), form a micellar solution from which lipid absorption takes place. After absorption into mucosal cells of the small intestine, triglyceride is resynthesized and packaged with small amounts of cholesterol, phospholipid, and protein to form chylomicrons. These are secreted into the intestinal lymphatics from which they enter the thoracic duct and then the venous circulation.

The crucial apolipoprotein in the genesis of chylomicrons is apo B-48. In humans, apo B-48 is formed only in the intestine as a result of an editing process in which the messenger ribonucleic acid (mRNA) for

TABLE 2.1 — LIPOPROTEIN CLASSES: COMPOSITION AND PARTICLE SIZE

	Chylomicrons	VLDL	LDL	HDL
Density (g/mL)	<1.006	<1.006	1.019-1.063	1.063-1.021
Electrophoretic mobility	(Origin)	Prebeta	Beta	Alpha-1
Size (diameter, nm)	75-300	30-80	20	7-10
% Composition				
Protein	1-2	10	25	50
Triglyceride	90-96	60	5	5
Cholesterol	2-5	12	50	20
Phospholipid	5	18	20	25
Major apoproteins	A-I,A-IV B-48 C-I, C-II, C-III E	B-100 C-I, C-II C-III, E	B-100	A-I, A-II C-I, C-II, C-III

Abbreviations: VLDL, very low-density lipoprotein; LDL, low-density lipoprotein; HDL, high-density lipoprotein.

a larger apo B molecule (apo B-100) is altered. The regulation of apo B secretion is largely posttranslational; much of the apo B that is synthesized is degraded intracellularly. The addition of triglyceride to apo B (catalyzed by the microsomal triglyceride transfer protein [MTP]) protects apo B from intracellular degradation. The disease abetalipoproteinemia is due to inability to produce MTP; these patients do not secrete any apo B-containing lipoproteins (chylomicrons, VLDL, and LDL) into the blood stream, and they develop deficiencies of fat-soluble vitamins.

In the circulation, chylomicrons rapidly accumulate apo E and apolipoproteins C-I, C-II, and C-III, which are transferred to them from HDL.

The triglyceride of chylomicrons is rapidly hydrolyzed by lipoprotein lipase (LPL), which resides on the surface of capillary endothelium. Apolipoprotein C-II serves as the activator of LPL. This generates smaller, shrunken particles, relatively richer in cholesterol and termed chylomicron remnants. These are rapidly removed from the circulation by the liver in a specific process involving the recognition of apo E. Additionally, in nonhepatic tissues, nonreceptor mechanisms may result in uptake of significant amounts of apo E-containing triglyceride-rich lipoproteins, such as remnants of chylomicron and VLDL (Al-Haideri et al, 1997). These mechanisms involve binding of apo E to heparan sulfate proteoglycans such as exist in arterial wall. This process is enhanced by the presence of LPL (Rutledge, 1997).

■ Genesis and Fate of VLDL

Very low-density lipoprotein is secreted from the liver as the major carrier in plasma of endogenously produced triglyceride. The major protein of VLDL is apo B-100. The secretion of apo B-100 is regulated largely by the availability of lipid as noted above for apo B-48 in the discussion of chylomicron metabolism. The C proteins and apo E are also important constituents of VLDL.

In a process analogous to that for chylomicrons, the triglyceride of VLDL undergoes LPL-mediated hydrolysis in capillary beds, and the surface components other than apo B are transferred to HDL. This produces a smaller, cholesterol-rich remnant particle. These remnants of VLDL are also termed intermediate-density lipoproteins (IDLs) since they fall between VLDL and LDL in density. In normal humans, approximately 10% to 30% of VLDL remnants are directly removed from the circulation by the liver or in nonhepatic tissues in a manner similar to the removal of chylomicron remnants. The remaining 70% to 90% of VLDL particles undergo further conversion to LDL. Apo B-100 is the sole protein constituent of LDL.

■ Genesis and Fate of LDL

Low-density lipoprotein is largely a catabolic product of VLDL. Nearly half of the body's pool of LDL is cleared from plasma daily. Two thirds of this clearance occurs via a receptor-mediated process involving recognition of apo B-100; this receptor-mediated pathway of clearance is termed the LDL receptor pathway, and the receptors are termed LDL receptors. (The editing process that produces apo B-48 of chylomicrons removes the portion of the apo B-100 molecule that is recognized by the LDL receptor.)

The activity of the LDL receptors is determined by nutritional, hormonal, and genetic factors as well as by cellular needs for cholesterol. When cholesterol is delivered to cells by any means, the activity of LDL receptors is suppressed. The actions of a sterol response element in the promoter region of the gene for the LDL receptor lead to this feedback regulation of LDL receptor activity. Additionally, the activity of these receptors is stimulated by both insulin and thyroid hormones.

Those LDL particles cleared by mechanisms other than the LDL receptor (normally about 15% of the plasma LDL pool daily) are removed by mechanisms including fluid-phase endocytosis and by type A "scavenger" receptors. The scavenger receptor pathway is important in foam cell formation and atherogenesis. The scavenger receptors recognize and bind chemically modified LDL (such as oxidized LDL) but not native LDL.

■ The Metabolism of HDL

High-density lipoprotein is produced by several mechanisms (Table 2.2). Both the liver and the intestine secrete discoidal HDL particles composed of apoproteins, phospholipid, and unesterified cholesterol. In addition to this direct secretion of HDL, it is produced as a result of lipolysis of VLDL and chylomicrons. In the lipolytic process, as the triglyceride-rich core of VLDL and chylomicrons is digested, a shrunken particle with a relative excess of polar surface matter (phospholipid, unesterified cholesterol, and apoproteins) is formed. The excessive surface material is excluded from the surface of the triglyceride-rich lipoproteins forming discoidal and vesicular "surface remnants" in the HDL density range. These discoidal particles may then either fuse with preex-

TABLE 2.2 — MECHANISMS INVOLVED IN FORMATION OF HIGH-DENSITY LIPOPROTEIN
• Secretion by liver and intestine
• Product of intravascular metabolism of very low-density lipoprotein and chylomicrons (surface remnants)
• Incorporation into preexisting high-density lipoprotein of cholesterol from cell membranes

isting spherical HDL particles or be acted upon by the enzyme lecithin-cholesterol acyltransferase (LCAT). LCAT converts unesterified cholesterol to cholesteryl esters, and this yields spherical particles with hydrophobic cholesteryl ester in the core. Furthermore, HDL may accumulate unesterified cholesterol from cell membranes. Cholesteryl ester in the core of HDL is removed by interaction with the cell surface scavenger receptor B1 (SR-B1). SR-B1 plays a crucial role in reverse cholesterol transport (transport of cholesterol from the periphery to the liver) and in provision of cholesterol to steroidogenic tissues. SR-B1 also facilitates efflux of cholesterol from nonhepatic cells to HDL, a crucial early step in reverse cholesterol transport (Jian et al, 1998; de la Llera-Moya, 1999).

■ Cholesteryl Ester Transfer Protein and Interactions Among Lipoprotein Classes

Cholesteryl ester transfer protein (CETP), a glycoprotein synthesized in the liver and secreted into the circulation, catalyzes the exchange of the nonpolar core components of lipoproteins (cholesteryl ester and triglyceride). The coordinated actions of CETP and LPL affect the concentrations and particle size of LDL and HDL. In patients with high levels of triglyceride-rich lipoproteins (VLDL and chylomicrons), the mass-action principle results in a CETP-mediated enrichment of LDL and HDL with triglyceride. Cholesteryl ester is concurrently lost from these particles by transfer to the triglyceride-rich lipoproteins. LPL can then hydrolyze the triglyceride which has been brought into LDL and HDL. The final result of this is a reduced concentration of LDL and HDL cholesterol (HDL-c); additionally, this process generates populations of LDL and HDL of reduced diameter and increased density. These structural changes in the LDL of hypertriglyceridemic patients result in defective interactions with receptors and defective metabolism of LDL.

Lipoproteins in Disease

Interest in the disorders of lipoprotein metabolism stems from the ability of the plasma lipoproteins to influence the development of atherosclerosis, xanthomatosis, and pancreatitis.

■ Lipoproteins and Atherogenesis

Macrophages, the precursors of foam cells of atherosclerotic lesions, are capable of storing massive amounts of cholesteryl esters in non–membrane-bound cytoplasmic droplets. The uptake of large amounts of cholesterol by these cells occurs via specific processes involving *type A scavenger receptors*. These receptors recognize and facilitate the uptake of chemically modified forms of LDL, but not native LDL. Oxidation is an important chemical modification of LDL, rendering it recognizable by scavenger receptors of macrophages.

Although scavenger receptors appear to play a predominant role in the formation of foam cells, LDL receptors may also contribute. The LDL receptors of macrophages are sluggishly down-regulated by plasma lipoproteins. Beta VLDL, a lipoprotein found in the plasma of subjects with type III hyperlipoproteinemia, is taken up by macrophages via the LDL receptor pathway.

Additionally, as noted earlier, remnants of triglyceride-rich lipoproteins can be taken up by nonreceptor mechanisms involving the binding to heparan sulfate proteoglycans.

Oxidized LDL is present in atherosclerotic lesions, but it is not present in a normal vessel wall. Oxidation of LDL may be atherogenic in multiple ways in addition to its being bound and internalized by scavenger receptors of macrophages, as reviewed in detail by Witztum and Steinberg (Table 2.3). Oxidized LDL is chemotactic for circulating monocytes. Thus when oxidized LDL enters arterial endothelium, it facilitates the accumulation of foam cell precursors. Furthermore, when these monocytes are transformed to macrophages, their motility is inhibited by oxidized LDL; thus they become more likely to stay within the vessel wall.

TABLE 2.3 — PROATHEROGENIC CHARACTERISTICS OF OXIDIZED LOW-DENSITY LIPOPROTEIN

- Uptake via macrophage scavenger receptors leading to accumulation of cholesteryl ester
- Chemotaxis for peripheral blood monocytes
- Inhibition of motility of tissue macrophages
- Cytotoxicity
- Stimulation of synthesis of monocyte chemotactic protein I by endothelial cells and smooth muscle cells
- Immunogenicity
- Procoagulant and platelet aggregatory activity
- Reduced responsiveness to nitric oxide (NO)-induced vasodilatation

Adapted from Witzum JL and Steinberg D. *J Clin Invest.* 1991; 88:1785-1792.

Another deleterious effect of oxidized LDL is that it appears to deplete stores of nitric oxide, which has been shown to be identical to endothelial-derived relaxing factor. Reduced levels of oxidized LDL may be the explanation for the reported beneficial effects of LDL-reducing therapy on endothelium-mediated vasodilatation in patients with coronary atherosclerosis (Treasure et al, 1995; Anderson et al, 1995).

Lp(a) is a lipoprotein generally present in low concentrations in the blood stream, but whose level in plasma has been correlated with the risk of CHD in many studies. The risk associated with elevated Lp(a) levels appears to depend to a great extent on the concurrent elevation of LDL levels. The unique protein of Lp(a) is termed apo(a) and is bound to apo B-100 by disulfide bonds. Apo(a) has considerable size heterogeneity among individuals, and there is a strong inverse relationship between the molecular weight of apo(a) and the plasma concentration of Lp(a). The amino acid sequence of apo(a) is strikingly homologous to that of plasminogen, and it appears in some circumstances to interfere with fibrinolysis. Lp(a) levels appear not to be affected by dietary manipulations or by the major lipid-lowering medications other than nicotinic acid. Treatment with estrogens also reduces Lp(a) levels. High levels of Lp(a) are associated with impaired arterial vasodilator capacity.

Small, dense LDL particles may have increased atherogenic potential. LDL particles of this sort can be produced, as noted above, by the coordinate activity of CETP and LPL. Several characteristics of these particles may make them particularly atherogenic (Table 2.4). They have greater propensity to be oxidized than do larger LDL particles. Additionally, conformational changes in apo B-100 in the small, dense LDL lead to defective interaction with receptors and

TABLE 2.4 — SMALL, DENSE LOW-DENSITY LIPOPROTEIN AND CORONARY HEART DISEASE RISK: POTENTIAL MECHANISMS RESPONSIBLE FOR AN ASSOCIATION
• Increased susceptibility to oxidation • Reduced affinity for the low-density lipoprotein receptor • Association with increased levels of very low-density lipoprotein remnants and low high-density lipoprotein

defective metabolism. Impaired capacity for clearance via the benign LDL receptor pathway increases clearance via scavenger receptors with consequent formation of foam cells.

A predominance of small, dense LDL particles is associated with a variety of other metabolic abnormalities (eg, low HDL, hypertriglyceridemia, obesity, insulin resistance, diabetes), which themselves are risk factors for premature atherosclerosis. It is not yet clear whether increased CHD risk is caused by the small, dense LDL particles themselves or by other associated abnormalities (Coresh and Kwiterovich, 1996).

Increased levels of HDL reduce the risk of CHD. Several mechanisms have been proposed to explain the inverse relationship between HDL levels and CHD risk (Table 2.5). HDL plays a role in reverse cholesterol transport, the transport of cholesterol from peripheral tissues to the liver. HDL is known to fuse with the polar "surface remnants" released from triglyceride-rich lipoproteins as they undergo lipolysis in plasma; these surface remnants may be harmful to vascular endothelium. By stimulating the synthesis

TABLE 2.5 — HDL AND CHD RISK: POTENTIAL MECHANISMS RESPONSIBLE FOR THE INVERSE RELATIONSHIP
• HDL fosters reverse cholesterol transport • HDL scavenges polar surface remnants released during lipolysis of VLDL and chylomicrons • HDL stimulates prostacyclin synthesis by arterial wall • HDL protects LDL from oxidation • Low HDL levels may be a marker for atherogenic abnormalities in another class of lipoproteins (eg, small, dense LDL, increased VLDL remnants)
Abbreviations: HDL, high-density lipoprotein; CHD, coronary heart disease; VLDL, very low-density lipoprotein; LDL, low-density lipoprotein.

of prostacyclin, HDL helps to prevent vasoconstriction. HDL has been shown to protect LDL from oxidation. Finally, low HDL levels may be a marker for abnormalities in other classes of lipoproteins (eg, high levels of VLDL remnants, high levels of small, dense LDL), which may themselves be atherogenic.

Uncommonly, HDL is increased as a result of deficiency of CETP. This mechanism for increasing HDL interferes with reverse cholesterol transport and *increases* coronary risk (Bruce et al, 1998). Similarly, in transgenic animal models, a deficiency of SR-B1 causes high HDL levels, impaired reverse cholesterol transport, and *increased* atherosclerosis. These, however, are the exceptions to prove the rule. In populations, high HDL levels can be assumed to be protective.

■ Lipoproteins and Pancreatitis

Abdominal pain and pancreatitis can complicate severe hypertriglyceridemia; this may occur when plasma triglyceride levels are above 1000 mg/dL. Since high concentrations of chylomicrons increase the viscosity of plasma, it may be speculated that marked chylomicronemia might lead to small, local areas of ischemia within the pancreas, thereby triggering the release of pancreatic lipase from cells; this lipase then catalyzes hydrolysis of the large amounts of triglyceride present in plasma, releasing large amounts of free fatty acids. When fatty acids are present in excess of the binding capacity of albumin, their detergent effect causes the lysis of cell membranes. In the pancreas, this leads to the release of additional amounts of lipase, and a chain reaction is initiated: production of fatty acids leading to cell lysis and release of lipase leading to more production of fatty acids, etc.

In the National Institutes of Health series of patients with the type V hyperlipoproteinemia phenotype (elevated VLDL and chylomicrons), pancreatitis was documented in 8 of 32 propositi and was probable in an additional 4 propositi (Greenberg et al, 1977). However, pancreatitis was uncommon in the type V relatives of the propositi, occurring in only 1 of 29. Pancreatitis occurs in most patients with type I hyperlipoproteinemia (chylomicronemia due to LPL deficiency), where triglyceride levels are even higher. Hemorrhagic pancreatitis is the major life-threatening risk for patients with type I hyperlipoproteinemia (chylomicronemia due to LPL deficiency).

Pathogenesis of Primary Elevations of LDL

Four classes of genetic conditions lead to elevation of LDL levels (Table 2.6): familial hypercholesterolemia, familial combined hyperlipidemia, familial defective apo B-100, and severe polygenic primary

TABLE 2.6 — GENETIC CONDITIONS CAUSING ELEVATED LOW-DENSITY LIPOPROTEIN

- *Familial Hypercholesterolemia:* Deficient or defective low-density lipoprotein (LDL) receptors; retarded clearance of LDL from plasma
- *Familial Combined Hyperlipidemia:* Increased secretion of apo B-100
- *Familial Defective apo B-100:* Mutant apo B-100 poorly recognized by LDL receptor; retarded clearance of LDL from plasma
- *Severe Primary Polygenic Elevation of LDL:* heterogeneous group of conditions; clearance of LDL usually retarded; E4 allele of apo-E sometimes plays a role

elevation of LDL. The population distribution of LDL is shown in Table 2.7 to provide a frame of reference.

■ Familial Hypercholesterolemia

This condition is caused by a genetic mutation leading to deficient or defective LDL receptors. As a consequence, the clearance of LDL from plasma is retarded, and LDL accumulates in plasma. Brown and Goldstein (1998) have categorized the genetic abnormalities of the LDL receptor as:

- Those which produce no receptors
- Those in which receptors are produced but do not move to the cell surface
- Those in which receptors have defective LDL-binding characteristics
- Internalization defects in which receptors locate on the cell surface but are not concentrated in the coated pits and are, therefore, not moved intracellularly with their bound LDL.

■ Familial Combined Hyperlipidemia

Familial combined hyperlipidemia is characterized by the presence of varying patterns of elevation of LDL and VLDL levels in families. Total plasma apo B levels tend to be high. This condition was first identified in a study of 500 survivors of myocardial infarction (Goldstein et al, 1973); it was found in 9.4% of survivors of myocardial infarction.

The pathogenesis of familial combined hyperlipidemia usually involves an increased rate of secretion of apo B-100 in VLDL and perhaps also direct hepatic secretion of LDL. However, in at least one kindred, impaired clearance is the main defect (Aguilar-Salinas, 1997). The precise mechanism responsible for the increased rate of secretion of lipoproteins containing apo B-100 remains unknown. Genetic abnormality within the apo A-I/C-III/A-IV gene locus seems most important in the genesis of familial combined hyperlipidemia (Dallinga-Thie et al, 1996). Abnormalities in the MTP may be involved. Some data have also suggested that abnormality in hormone-sensitive lipase (responsible for hydrolyzing triglyceride in adipocytes) may be involved in the pathogenesis of familial combined hyperlipidemia.

Depending on the efficiency of metabolic conversion of VLDL to LDL, increased secretion of VLDL apo B may lead to accumulation of high levels of either or both of these lipoproteins.

The mechanism responsible for increased risk of atherosclerosis in this disorder has not been fully defined. Possible explanations for the increased atherosclerosis associated with familial combined hyperlipidemia include transient or long-lasting elevation of LDL, qualitative abnormality of LDL (small, dense LDL), qualitative or quantitative abnormalities of HDL, or qualitative abnormalities of VLDL.

Familial combined hyperlipidemia has been associated with the presence of small, dense LDL and of familial hyperapobetalipoproteinemia (increased apo B-100 levels with normal LDL cholesterol), conditions which also seem to be atherogenic.

■ Familial Defective apo B-100

This condition is caused by mutation in the gene for apo B-100, causing LDL to bind poorly to receptors. Thus this condition is functionally similar to familial hypercholesterolemia, in which the receptors are defective.

Two different mutations have been shown to produce familial defective apo B-100. The population frequency of this disorder is approximately 1/500, similar to that of familial hypercholesterolemia. The LDL of patients with familial defective apo B-100 appears to be particularly prone to oxidation, and this probably contributes to their increased CHD risk.

TABLE 2.7 — DISTRIBUTION OF LDL-C (MG/DL) AMONG ADULTS (≥20 YEARS OF AGE) IN THE UNITED STATES

Age (Years)	Mean	Mean Percentiles of LDL								
		5TH	10TH	15TH	25TH	50TH	75TH	85TH	90TH	95TH
Men										
20-34	120	67	78	86	97	121	139	152	165	186
35-44	134	85	92	98	111	131	156	166	176	192
45-54	138	78	91	100	118	136	163	174	187	195
55-64	142	78	90	104	117	143	165	175	194	205
65-74	141	93	104	109	119	134	163	177	185	199
≥75	132	83	88	93	106	130	154	170	186	196

Women										
20-34	110	59	70	75	88	108	129	142	155	173
35-44	117	67	85	88	97	116	138	146	155	165
45-54	132	70	87	93	107	130	157	173	182	198
55-64	145	79	90	101	122	145	170	184	189	209
65-74	147	92	97	109	119	148	169	185	192	206
≥75	147	90	102	109	121	143	168	189	197	209

Abbreviation: LDL-c, low-density lipoprotein cholesterol.

Data taken from National Health and Nutrition Survey III (NHANES III), 1988 through 1991. Table adapted from National Cholesterol Education Program. *Circulation*. 1994;89:1333-1445.

■ **Severe Primary Polygenic Elevation of LDL**

Severe primary polygenic elevation of LDL is relatively common, being far more prevalent than the known specific single-gene disorders. These patients probably have a heterogeneous group of conditions with varying pathophysiology. In most patients with polygenic hypercholesterolemia, clearance of LDL is retarded and the lipoproteins are qualitatively normal.

One genetic trait causing modest elevation of LDL-c is presence of the E4 allele of apo E. Additionally, the expression of this allele appears to be an independent coronary risk factor.

The most common cause of moderate elevation of LDL in the North American and European populations is not a primary genetic disorder; rather, it is dietary excess of saturated fats and cholesterol.

Elevation of IDL: Type III Hyperlipoproteinemia (Familial Dysbetalipoproteinemia)

Type III hyperlipoproteinemia (Table 2.8) is characterized by an accumulation of remnants of VLDL and chylomicrons and the premature development of atherosclerotic disease.

In type III hyperlipoproteinemia, the accumulation of remnants causes elevation of both cholesterol

TABLE 2.8 — FAMILIAL TYPE III HYPERLIPOPROTEINEMIA (FAMILIAL DYSBETALIPOPROTEINEMIA)
• *Pathogenesis:* Defective apolipoprotein E (usually apo E-II/E-II phenotype) • *Lipoproteins:* Hypercholesterolemia and hypertriglyceridemia; elevated remnants of very low-density lipoprotein and chylomicrons

and triglyceride concentrations in plasma, usually in the range of 250 to 500 mg/dL.

An abnormality of apo E is fundamental for the expression of this condition. Several different mutations of apo E which interfere with its interaction with cell-surface receptors can lead to type III hyperlipoproteinemia when present in homozygous form; these generally lead to the apo E-II/E-II phenotype. At least one mutation leads to dominant expression of type III hyperlipoproteinemia. Absence of apo E can also lead to type III hyperlipoproteinemia. The absence of a functionally normal apo E leads to impaired clearance of remnant particles from plasma.

The commonest apo E mutation associated with type III hyperlipoproteinemia (apo E-II, arg158 → cys) has a gene frequency of 10%. Homozygosity, therefore, occurs in approximately 1% of the population. However, the frequency of type III hyperlipoproteinemia is much lower (0.01% to 0.04%). Thus, the abnormality of apo E is a necessary condition, but not a sufficient condition, for the occurrence of type III hyperlipoproteinemia. Conditions which increase the synthesis of VLDL (obesity, caloric excess, and alcohol use) often precipitate type III hyperlipoproteinemia in individuals who are predisposed to this condition. Here, the burden of increased rates of synthesis apparently can overwhelm marginal clearance mechanisms. Hypothyroidism can also precipitate type III hyperlipoproteinemia in patients with defective apo E since the hypothyroid state suppresses the synthesis of hepatic LDL receptors, further impairing clearance mechanisms.

A very rare cause of accumulation of remnant lipoproteins (the type III phenotype) in patients with normal apo E is *deficiency of hepatic triglyceride lipase*, an enzyme involved in conversion of VLDL remnants to LDL. HDL levels are generally elevated in these patients; in typical type III patients, HDL levels are quite low.

Hypertriglyceridemias

The role of hypertriglyceridemia as an independent risk factor remains controversial. A meta-analysis of studies including 22,499 men and 6345 women indicates that a triglyceride elevation of 89 mg/dL (1 mM) is associated with a 14% increase in coronary risk in men and a 37% increase in risk in women (after correction for associations with HDL cholesterol and other risk factors) (Austin, 1999). However, a sophisticated analysis of data from three large prospective studies involving 15,880 subjects indicated that measurement of fasting serum triglyceride levels does not provide an independent contribution [of LDL and HDL cholesterol] to the assessment of CHD risk (Avins and Neuhaus, 2000).

The hypertriglyceridemias have been categorized by the National Cholesterol Education Program as borderline (triglyceride levels 200 to 400 mg/dL), high triglycerides (400 to 1000 mg/dL), and very high triglycerides (>1000 mg/dL). This categorization is therapeutically useful, but it provides little insight into pathophysiology. A heterogeneous group of conditions are responsible for the primary hypertriglyceridemias (Table 2.9).

Familial combined hyperlipidemia, whose pathophysiology was discussed earlier, is a cause of borderline-high triglyceride levels and of high triglyceride levels. This condition increases risk of CHD.

Lipoprotein lipase deficiency is a rare cause of very high triglyceride levels. This can be caused by two classes of mutations, those leading to:

- The production of inactive LPL molecules
- An absence of LPL molecules.

TABLE 2.9 — PATHOPHYSIOLOGIC MECHANISMS OF HYPERTRIGLYCERIDEMIAS	
Mechanism (Genetic Condition)	**Impact on Physiology**
Increased apo B-100 secretion (familial combined hyperlipidemia)	Increased number of very low-density lipoprotein (VLDL) particles
Increased hepatic triglyceride production (familial hypertriglyceridemia)	Increased size of VLDL particles
Increased expression of apo C-III or apo C-II	Reduced VLDL clearance
Heterozygosity for lipoprotein lipase deficiency (some cases of familial hypertriglyceridemia)	Reduced clearance of triglyceride-rich lipoproteins
Lipoprotein deficiency (rare)(type I hyperlipoproteinemia)	Reduced clearance of triglyceride-rich lipoproteins
Apolipoprotein C-II deficiency (rare)(type I or type V hyperlipoproteinemia)	Reduced clearance of triglyceride-rich lipoproteins

Many mutations leading to LPL deficiency have been identified. Triglyceride levels generally exceed 1000 mg/dL, and the triglyceride elevation is due largely to the persistence of chylomicrons in plasma. LDL and HDL levels tend to be very low; the low levels of LDL and HDL are a result of the action of CETP in the presence of high levels of triglyceride-rich lipoproteins (*vide supra*). This condition does not generally increase CHD risk.

Deficiency of apo C-II, the activator of LPL, is another rare cause of very high triglyceride levels.

Familial hypertriglyceridemia refers to a group of conditions causing borderline-high and high triglyceride levels. In most cases, these conditions are due primarily to excessive production of triglyceride. VLDL apo B-100 secretion rates tend to be normal, and hepatic triglyceride secretion is high. Since a single apo B-100 molecule is present in each VLDL particle, the VLDL particles are large and triglyceride-rich. These large VLDL particles are probably not very atherogenic. The high secretion rates of VLDL-triglyceride often saturate the clearance mechanisms (involving LPL) for triglyceride-rich lipoproteins.

In some families, heterozygosity for LPL deficiency is the cause of familial hypertriglyceridemia.

Expression in transgenic mice of human apo C-III or apo C-II, two of the apolipoproteins of VLDL, can cause hypertriglyceridemia by retarding the clearance of VLDL in these animals. There is evidence that abnormalities in the expression of the apo C-II gene or the apo C-III gene may sometimes be related to hypertriglyceridemia in humans.

Pathogenesis of Hypoalphalipoproteinemia

Familial hypoalphalipoproteinemia, causing very low levels of HDL, has recently been demonstrated to be caused by mutation of the ABC1 gene. This gene codes for the ABC1 protein. This protein plays a critical role in reverse cholesterol transport. In macrophages and other reticuloendothelial cells, it appears to be involved in the hand-off of cellular cholesterol to HDL. Defects of this gene are also responsible for Tangier disease, in which there is a near absence of HDL (Hobbs and Rader, 1999).

High triglyceride levels cause a reduction in HDL-c levels by a mechanism involving the CETP as noted.

The population distribution of HDL-c is shown in Table 2.10.

SELECTED READINGS

Al-Haideri M, Goldberg IJ, Galeano NF, et al. Heparan sulfate proteoglycan-mediated uptake of apolipoprotein E-triglyceride-rich lipoprotein particles: a major pathway at physiological particle concentrations. *Biochemistry*. 1997;36:12766-12772.

Aalto-Setala K, Fisher EA, Chen X, et al. Mechanism of hypertriglyceridemia in human apolipoprotein (apo) CIII transgenic mice. Diminished very low-density lipoprotein fractional catabolic rate associated with increased apo CIII and reduced apo E on the particles. *J Clin Invest*. 1992;90:1889-1900.

Aguilar-Salinas CA, Hugh P, Barrett R, Pulai J, Zhu XL, Schonfeld G. A familial combined hyperlipidemic kindred with impaired apolipoprotein B catabolism. Kinetics of apolipoprotein B during placebo and pravastatin therapy. *Arterioscler Thromb Vasc Biol.* 1997;17:72-82.

Avins AL, Neuhaus JM. Do triglycerides provide meaningful information about heart disease risk? *Arch Intern Med.* 2000;160: 1937-1944.

Anderson TJ, Meredith IT, Yeung AC, Frei B, Selwyn AP, Ganz P. The effect of cholesterol-lowering and antioxidant therapy on endothelium-dependent coronary vasomotion. *N Engl J Med.* 1995;332:488-493.

Austin MA. Epidemiology of hypertriglyceridemia and cardiovascular disease. *Am J Cardiol.* 1999;83:13F-16F.

Breslow JL. Lipoprotein transport gene abnormalities underlying coronary heart disease susceptibility. *Annu Rev Med.* 1991;42:357-371.

Brooks-Wilson A, Marcil M, Clee SM, et al. Mutations in ABC1 in Tangier disease and familial high-density lipoprotein deficiency. *Nat Genet*. 1999;22:336-345.

Brown MS, Goldstein JL. A receptor-mediated pathway for cholesterol homeostasis. *Science*. 1986;232:34-47.

TABLE 2.10 — DISTRIBUTION OF HDL-C (MG/DL) AMONG ADULTS (≥20 YEARS OF AGE) IN THE UNITED STATES

Age (Years)	Mean	Mean Percentiles of HDL								
		5TH	10TH	15TH	25TH	50TH	75TH	85TH	90TH	95TH
Men										
20-34	47	30	34	35	38	46	54	60	64	71
35-44	46	28	30	33	37	44	53	58	63	73
45-54	47	38	30	33	36	43	53	61	66	77
55-64	46	29	31	33	36	43	53	59	62	72
65-74	45	28	31	32	36	43	53	58	62	71
≥75	47	28	32	34	38	45	54	62	67	75

Women										
20-34	56	34	38	41	44	54	64	70	75	83
35-44	54	33	37	40	44	53	64	69	72	79
45-54	57	37	38	41	46	56	65	72	77	84
55-64	56	33	37	40	44	53	66	73	79	87
65-74	56	34	37	40	44	54	65	73	78	87
≥75	57	33	39	41	44	56	66	73	78	87

Abbreviation: HDL-c, high-density lipoprotein cholesterol.

Data taken from National Health and Nutrition Survey III (NHANES III), 1988 through 1991. Table adapted from National Cholesterol Education Program. *Circulation*. 1994;89:1333-1445.

Bruce C, Chouinard RA Jr, Tall AR. Plasma lipid transfer proteins, high-density lipoproteins, and reverse cholesterol transport. *Annu Rev Nutr*. 1998;18:297-330.

Coresh J, Kwiterovich PO Jr. Small, dense low-density lipoprotein particles and coronary heart disease risk. A clear association with uncertain implications. *JAMA*. 1996;276:914-915.

Dallinga-Thie GM, Bu XD, van Linde-Sibenius Trip M, Rotter JI, Lusis AJ, de Bruin TW. Apolipoprotein A-I/C-III/A-IV gene cluster in familial combined hyperlipidemia: effects on LDL-cholesterol and apolipoproteins B and C-III. *J Lipid Res*. 1996;37:136-147.

de la Llera-Moya M, Rothblat GH, Connelly MA, et al. Scavenger receptor BI (SR-BI) mediates free cholesterol flux independently of HDL tethering to the cell surface. *J Lipid Res*. 1999;40:575-580.

Goldstein JL, Schrott HG, Hazzard WR, Bierman EL, Motulsky AG. Hyperlipidemia in coronary heart disease. II. Genetic analysis of lipid levels in 176 families and delineation of a new inherited disorder, combined hyperlipidemia. *J Clin Invest*. 1973;52:1544-1568.

Greenberg BH, Blackwelder WC, Levy RI. Primary type V hyperlipoproteinemia. A descriptive study in 32 families. *Ann Intern Med*. 1977;87:526-534.

Hobbs HH, Rader DJ. ABC1: connecting yellow tonsils, neuropathy, and very low HDL. *J Clin Invest*. 1999;104:1015-1017.

Innerarity TL, Mahley RW, Weisgraber KH, et al. Familial defective apolipoprotein B-100: a mutation of apolipoprotein B that causes hypercholesterolemia. *J Lipid Res*. 1990;31:1337-1349.

Jian B, de la Llera-Moya M, Ji Y, et al. Scavenger receptor class B type I as a mediator of cellular cholesterol efflux to lipoproteins and phospholipid acceptors. *J Biol Chem*. 1998;273:5599-5606.

Lawn RM, Wade DP, Garvin MR, et al. The Tangier disease gene product ABC1 controls the cellular apolipoprotein-mediated lipid removal pathway. *J Clin Invest*. 1999;104:R25-R31.

Maher VM, Brown BG, Marcovina SM, Hillger LA, Zhao XQ, Albers JJ. Effects of lowering elevated LDL cholesterol on the cardiovascular risk of lipoprotein (a). *JAMA*. 1995;274:1771-1774.

Marcil M, Brooks-Wilson A, Clee SM, et al. Mutations in the ABC1 gene in familial HDL deficiency with defective cholesterol efflux. *Lancet*. 1999;354:1341-1346.

Myant NB. Familial defective apolipoprotein B-100: a review, including some comparisons with familial hypercholesterolemia [published erratum appears in *Atherosclerosis*. 1994;105:253]. *Atherosclerosis*. 1993;104:1-18.

National Cholesterol Education Program. Second Report of the Expert Panel on Detection, Evaluation, and Treatment of High Blood Cholesterol in Adults (Adult Treatment Panel II). *Circulation*. 1994;89:1333-1445.

Rutledge JC, Woo MM, Rezai AA, Curtiss LK, Goldberg IJ. Lipoprotein lipase increases lipoprotein binding to the artery wall and increases endothelial layer permeability by formation of lipolysis products. *Circ Res*. 1997;80:819-828.

Schaefer EJ, Lamon-Fava S, Jenner JL, et al. Lipoprotein (a) levels and risk of coronary heart disease in men. The Lipid Research Clinics Coronary Primary Prevention Trial. *JAMA*. 1994;271:999-1003.

Treasure CB, Klein JL, Weintraub W, et al. Beneficial effects of cholesterol-lowering therapy on the coronary endothelium in patients with coronary artery disease. *N Engl J Med*. 1995;332:481-487.

Wilson PW, Myers RH, Larson MG, Ordovas JM, Wolf PA, Schaefer EJ. Apolipoprotein E alleles, dyslipidemia, and coronary heart disease. The Framingham Offspring Study. *JAMA*. 1994;272:1666-1671.

Witztum JL, Steinberg D. Role of oxidized low density lipoprotein in atherogenesis. *J Clin Invest*. 1991;88:1785-1792.

3 Classification of Familial Hyperlipidemia

While the public focus of the National Cholesterol Education Program (NCEP) report was on high cholesterol, contemporary practice has moved beyond using only lipid values for both risk determination and treatment. For screening purposes, nonfasting values for cholesterol and high-density lipoprotein cholesterol (HDL-c) can be used. The error involved in using nonfasting values for screening is not great in most circumstances. This practical tip is of greatest use in young patients without known risk factors for coronary heart disease (CHD). Here nonfasting values of total cholesterol under 200 mg/dL and HDL-c over 50 mg/dL indicate low-risk subjects in whom further lipid testing is not likely to be informative. On the other hand, for those with multiple risk factors, a family history of premature CHD, or overt atherosclerotic vascular disease, the first test should be a fasting lipid profile which includes the total cholesterol, triglycerides (TG), HDL-c, and low-density lipoprotein cholesterol (LDL-c). The LDL-c is calculated by the following formula, known as the Friedewald formula:

$$\text{LDL-c} = \text{Total cholesterol} - \text{HDL-c} - (\text{TG}/5)$$

For example, if total cholesterol is 300, triglycerides are 250, and HDL-c is 50, the formula determines LDL-c as:

$$\text{LDL-c} = 300 - 50 - (250/5);$$
$$\text{LDL-c} = 200 \text{ mg/dL by calculation.}$$

The formula should not be used when triglycerides exceed 400 mg/dL or the cholesterol-rich, very low-density lipoproteins (VLDL) seen in familial dysbetalipoproteinemia are present. Fortunately, this latter condition is very rare (see Chapter 5, *Clinical Appraisal and Goals of Therapy,* for use of formula in screening).

Another formula has gained attention in the past few years and is worthy of mention. It is used in the calculation of non-HDL cholesterol. By definition, it includes the cholesterol content of the apo B-containing lipoproteins LDL and VLDL. It has the advantage of not being influenced by increasing triglyceride values, which serve to lower the LDL-c calculated by formula. The non–HDL-c is simple to calculate and in one study of normolipidemic subjects, it correlated well with apo B (Abate et al, 1993):

$$\text{Non-HDL-c} = \text{Total cholesterol} - \text{HDL-c}$$

Direct measurement of LDL cholesterol offers advantages of greater accuracy when triglycerides are in the 200 to 400 mg/dL range. This information may be particularly useful in the follow-up of diabetic patients who often have high triglycerides as well as elevated LDL-c (Hirany et al, 1997).

A useful classification of hypertriglyceridemia has been proposed by the NCEP. Borderline-high triglyceride levels are 200 to 400 mg/dL and very high levels are 1000 mg/dL or higher. To confirm fasting chylomicrons, visual inspection of an upright tube of plasma after it is chilled to 4°C is suggested. A creamy supernatant layer indicates chylomicrons are present (Figure 3.1).

In the mid-1960s, the Typing System was popularized by Dr. Donald Fredrickson and his co-workers at the National Institutes of Health. This system was an important advance because it focused interest

FIGURE 3.1 — CLASSIFICATION OF HYPERTRIGLYCERIDEMIA: PRESENCE OF CHYLOMICRONS IN PLASMA

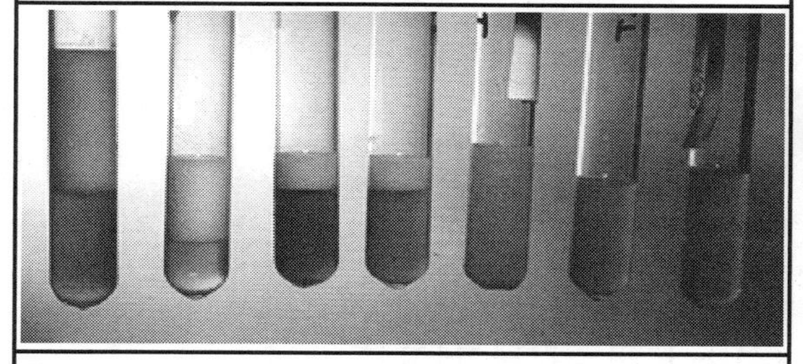

Presence of chylomicrons is visually noted (from left to right) by the appearance of a creamy, supernatant layer following 4°C chilling of test tube plasma sample. These tubes reflect daily improvements in chylomicronemia in a patient recovering from pancreatitis.

away from cholesterol and triglyceride onto the various lipoprotein classes. Knowledge of chylomicrons, LDL, VLDL, and HDL did bring increased complexity to the field of lipid metabolism.

Table 3.1 lists the lipoprotein types, using the familiar rubrics of I through V. Lipoprotein electrophoresis is a useful way to show the qualitative differences that highlight the differences between these types. Here LDL excess stains in the beta region; VLDL excess, in the pre-beta region; and HDL in the alpha region of the electrophoretic strip. Hence abeta-, hypobeta-, or hyperbetalipoproteinemia describes differences in LDL. Likewise, hyperalpha- or hypoalphalipoproteinemia describes differences in HDL.

However, while still a useful shorthand to describe the kind of lipoprotein excess seen in an individual patient, the electrophoretically determined phenotypes are not used today for two compelling reasons. First, they do not convey quantitative information. Second, the types do not convey precise ge-

TABLE 3.1 — LIPOPROTEIN TYPES BASED ON LIPOPROTEIN ELECTROPHORETIC PATTERNS (NOT GENOTYPES)

Type	Electrophoretic marker	Significance
Type I	Chylo band at the origin	Fasting chylomicronemia
Type II	Increased beta band	Elevated LDL-c
Type IIa	Increased beta band	Elevated LDL-c; normal triglycerides
Type IIb	Increased beta, prebeta band	Elevated LDL-c; elevated triglycerides
Type III	Floating beta band	Abnormal VLDL
Type IV	Increased prebeta band	Increased VLDL-c
Type V	Increased prebeta, chylo band	Fasting chylomicronemia and VLDL-c

Abbreviations: LDL, low-density lipoprotein; VLDL, very low-density lipoprotein; VLDL-c, very low-density lipoprotein cholesterol.

netic information. For example, among kindred with familial combined hyperlipidemia, described in survivors of myocardial infarction (MI) in Seattle, multiple lipoprotein phenotypes (IIa, IIb, IV) could be seen. Thus the typing system did fulfill its mission of stimulating interest in the genetic and metabolic differences of lipid disorders, but it has given way to more quantitative techniques.

The genetics of lipid disorders are considered in detail in Chapter 2, *Pathophysiology of Hyperlipoproteinemias*. A clinical overview of disorders seen in practice is given in Table 3.2. They are organized into four categories:

- Disorders of predominant excess LDL-c
- Disorders with mild to moderate hypertriglyceridemia
- Disorders involving low levels of HDL-c
- Disorders involving elevated levels of Lp(a)
- Disorders involving severe hypertriglyceridemia.

The practicing physician is urged to review these disorders carefully for several reasons. Accurate genetic diagnosis is often essential for effective treatment. Suspicion of genetic disorders leads to family screening and identification of an even greater number of affected relatives. This identification may allow therapy to be started before clinical problems such as CHD are apparent.

> ***Genetic Tip****:* When a patient with a CHD event is identified as having a lipid disorder, screening family members must be considered as important as ruling out secondary causes and prescribing therapy.

The most common genetic disorders seen in practice are listed in Tables 3.2 and 3.3. Even though the

TABLE 3.2 — COMMON GENETIC DISORDERS SEEN IN PRACTICE (RANKED BY PREDOMINANT LIPID ABNORMALITY)

- High low-density lipoprotein (LDL) cholesterol
  - Familial hypercholesterolemia
  - Familial defective apo B
  - Familial combined hyperlipidemia
  - Polygenic primary elevation of LDL
- High triglyceride of mild to moderate severity
  - Familial combined hyperlipidemia (FCHL)
  - Familial dysbetalipoproteinemia, also known as familial type III
  - Familial hypertriglyceridemia
- Low high-density lipoprotein (HDL)
  - Familial hypoalphalipoproteinemia
- Excess Lp(a)
- High triglycerides—severe (chylomicronemia syndrome)
  - Familial lipoprotein lipase deficiency

TABLE 3.3 — POPULATION-BASED FREQUENCY OF DYSLIPIDEMIA SYNDROMES IN CORONARY-PRONE FAMILIES IN UTAH

Familial combined hyperlipidemia	36% to 48% of families with CHD
Familial dyslipidemic hypertension	21% to 54% of families with CHD
Isolated low levels of HDL-c	15%
Familial hypercholesterolemia	3%
Familial type III	3%
No recognizable lipid abnormalities	15%

Abbreviations: CHD, coronary heart disease; HDL-c, high-density lipoprotein cholesterol.

Williams RR, et al. *Arch Intern Med.* 1990;150:582-588.

cases have been documented in different manners, they are presented together.

> ***Genetic Tip***: Notice how frequently deficiencies of HDL-c are seen in cases of premature CHD.

Lipid Disorders in Familial CHD

In the first study, Williams and co-workers determined the frequency of familial lipid abnormalities in 33 families where two or more siblings had CHD by the age of 55. They found that 75% of the persons with early CHD in these families had either cholesterol or triglyceride above the 90th percentile and/or HDL-c below the 10th percentile. Williams observed the following in these high-risk, coronary-prone families:

- HDL-c and triglyceride abnormalities were twice as common as LDL-c abnormalities.
- Low HDL-c alone was seen five times more often than the monogenic syndromes of familial hypercholesterolemia and familial type III.

Lipid Disorders in Premature CHD

A more detailed lipoprotein analysis was carried out by Dr. Jacques Genest and co-workers from the laboratory of Dr. Ernst Schaefer. His group investigated 102 families (603 subjects) in whom the proband had significant coronary artery disease (CAD) before the age of 60, documented by angiography (Table 3.4). They noted that among these patients with CAD:

- More than half had a familial lipid or apolipoprotein disorder.
- Low HDL-c was the most common disorder; it was seen in 39.2% of cases.
- Those with excess Lp(a) had the highest degree of parent-offspring correlation.

TABLE 3.4 — PATIENTS WITH CAD AT ANGIOGRAPHY BEFORE AGE 60	
Lp(a) excess (includes 12.7% with no other dyslipidemias)	18.6%
Hypertriglyceridemia with low HDL-c	14.7%
Combined hyperlipidemia with low HDL-c	11.7%
Elevated apo B only	5%
Low HDL-c	4%
Combined hyperlipidemia (↑ LDL + TG)	2%
Elevated LDL only	3%
High LDL-c with low HDL-c	2%
Hypertriglyceridemia only	1%
Decreased apo A-I only	1%

Low HDL-c = HDL below the tenth percentile; hypertriglyceridemia = triglycerides above the ninetieth percentile; hypercholesterolemia = LDL above ninetieth percentile.

Abbreviations: CAD, coronary artery disease; HDL-c, high-density lipoprotein cholesterol; LDL, low-density lipoprotein; TG, triglycerides; LDL-c, low-density lipoprotein cholesterol; HDL, high-density lipoprotein.

Genest J Jr, et al. *J Am Coll Cardiol.* 1992;19:792-802.

In addition, this group looked more intensively at apolipoprotein abnormalities in 321 men with coronary disease documented by angiography who were 50 years old on the average. They were compared with age-matched controls chosen from the Framingham Offspring Study who were clinically free of CHD (Figures 3.2 and 3.3). After correction for sampling in hospital, use of beta blockers, and effects of diet, there were substantial differences, with patients having higher values for total cholesterol, LDL-c, apo B, Lp(a), and triglycerides than controls. Also, there

FIGURE 3.2 — LIPIDS IN CORONARY ARTERY DISEASE PATIENTS VS CONTROLS

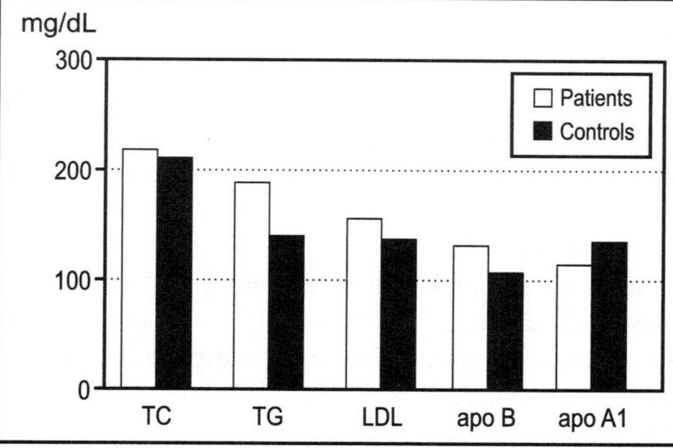

Abbreviations: TC, total cholesterol; TG, triglyceride; LDL, low-density lipoprotein; apo, apolipoprotein.

Genest J Jr, et al. *J Am Coll Cardiol*. 1992;19:792-802.

FIGURE 3.3 — HIGH-DENSITY LIPOPROTEIN CHOLESTEROL AND LP(A) IN CORONARY ARTERY DISEASE PATIENTS VS CONTROLS

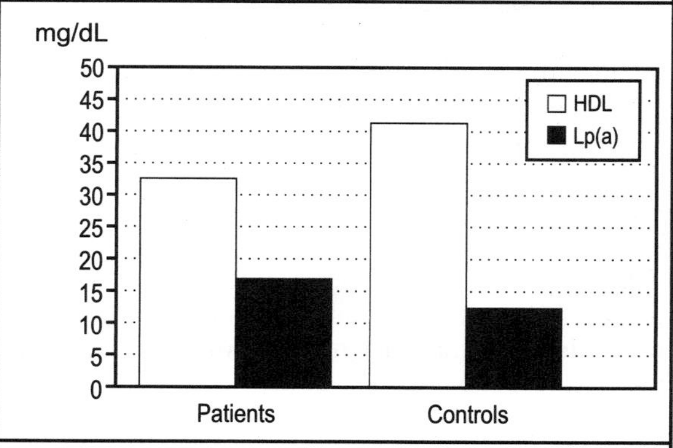

Abbreviation: HDL, high-density lipoprotein.

Genest J Jr, et al. *J Am Coll Cardiol*. 1992;19:792-802.

were lower values for HDL-c and apo A-I in patients as contrasted with controls.

Stepwise, discriminant analysis can determine those factors that contribute independently to risk. In this study, smoking, hypertension, decreased apo A-I, increased apo B, increased Lp(a), and diabetes were all significant ($P < 0.05$) factors in descending order of importance in distinguishing patients with coronary artery disease (CAD) from normal control subjects. These data provided support for measuring HDL-c as well as total cholesterol in screening efforts. The authors noted that 35% of patients had a total cholesterol level less than 200 mg/dL after correction; of those patients, 73% had an HDL-c level less than 35 mg/dL.

Familial Lipid Disorders

Those Disorders Characterized Predominantly by High Cholesterol Due to Excess LDL-c: Prototype Disorder—Familial Hypercholesterolemia

---

■ Familial Hypercholesterolemia

Familial hypercholesterolemia (FH) is inherited as an autosomal dominant disorder. Brown and Goldstein demonstrated through a series of pivotal experiments that patients with FH either lack or have defective receptors on hepatic cells (see Chapter 2, *Pathophysiology of Hyperlipoproteinemias*).

The estimated prevalence is 0.2% for heterozygous cases. Elevated cholesterol and LDL-c levels are present from birth. Secondary causes of marked hypercholesterolemia that must be distinguished from this disorder include:

- Hypothyroidism
- Obstructive liver disease
- Nephrotic syndrome.

Typical cholesterol values range from 325 to 450 mg/dL for heterozygous cases and 500 to 1000 mg/dL for homozygous cases. The clinical picture includes characteristic tendon xanthomas, which are present in over 70% of cases by age 30. These must be sought over Achilles tendons, extensor tendons over the metacarpals, and even hallucis longus tendons in the feet (Figure 3.4). Arcus corneae is not a sensitive marker of the condition, particularly when seen in older patients or in those of African descent. On the other hand, when seen in the inferior and superior poles of white patients under age 35, it should raise the suspicion of FH.

FIGURE 3.4 — TENDON XANTHOMAS IN FAMILIAL HYPERCHOLESTEROLEMIA

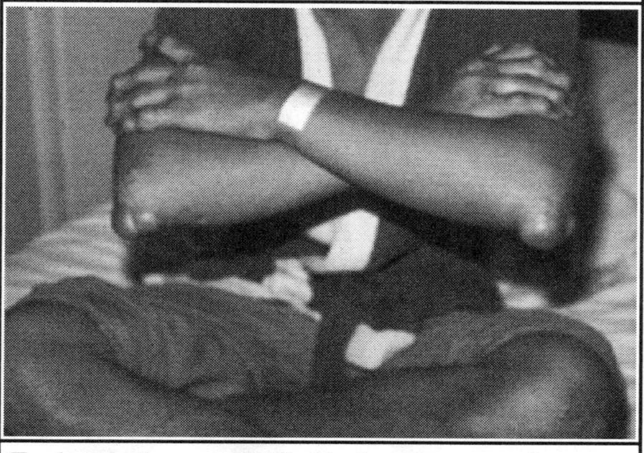

Tendon xanthomas exhibited in the extensor tendons over the metacarpals and elbows.

The most serious symptoms are sudden death, MI, or angina pectoris from atherosclerotic CAD. This occurs in the fourth to fifth decade of life in untreated subjects. Similarly affected women develop premature CHD as well, but the onset is delayed 10 years as compared with their male counterparts. Cigarette smoking, low HDL-c, and a family history of premature CHD are markers for those who are more likely to experience early coronary events. Some, but not all, studies suggest that increased Lp(a) is also an important marker of propensity for CHD.

The involvement of the coronary arteries in familial hyperlipidemia may be greater than previously assumed. In one study of asymptomatic, nonsmoking subjects with FH, intravascular ultrasound detected diffuse plaque in the left anterior descending artery and left main coronary artery despite minimal or no changes on the coronary angiogram (Figure 3.5). Levels of HDL-c correlated inversely and total/HDL-c ratios correlated directly with the extent and severity of the coronary plaque, documented by intravascular ultrasound (IVUS) imaging of the left main and left an-

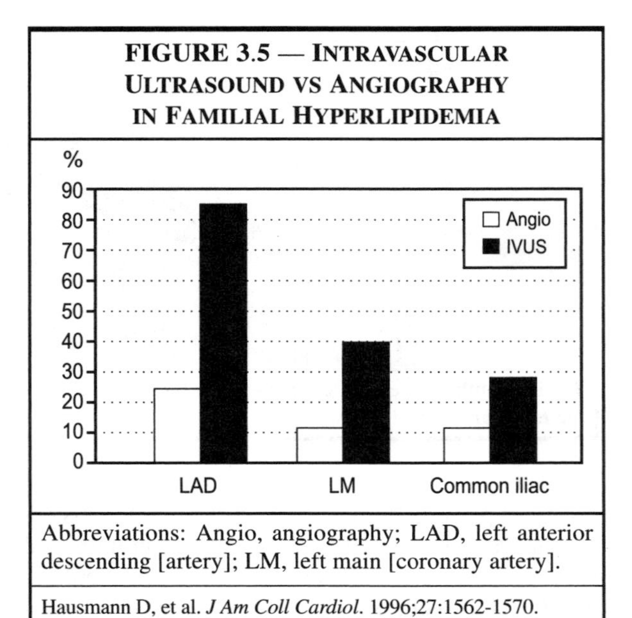

FIGURE 3.5 — INTRAVASCULAR ULTRASOUND VS ANGIOGRAPHY IN FAMILIAL HYPERLIPIDEMIA

Abbreviations: Angio, angiography; LAD, left anterior descending [artery]; LM, left main [coronary artery].

Hausmann D, et al. *J Am Coll Cardiol.* 1996;27:1562-1570.

terior descending arteries. A serial angiographic study compared intensive, multiple drug therapies with either diet or diet and resin therapy for those with FH. Overall, there was a mean change toward regression in the treated group, whereas the control group showed progression. Regression was seen among female patients with FH when this group was analyzed separately.

3

Homozygous cases are rare; they occur in approximately one of each million births. Xanthomas are seen during the first decade of life, and coronary disease is seen during childhood. Drug therapy is often ineffective. The consequence of not lowering the markedly elevated LDL-c levels is severe. The patient develops obstructive proximal coronary disease as well as aortic and supravalvular stenosis and often dies before age 20. Dramatic treatments, such as weekly apheresis, in which LDL is removed by running the blood through columns which remove apo B–containing lipoproteins, or liver transplantation, have provided means of controlling LDL-c and have made longer life spans possible.

An important concept in cholesterol screening was advanced by Medical Pedigrees With FH to Make Early Diagnoses and Prevent Early Deaths (MED PED FH). This Utah-based project noted that since the likelihood of finding a case of FH in a first-degree relative was 50% as compared with a 1-in-500 chance of finding an FH case in the general population, the screening criteria should be different. Evaluation suggested that at a cholesterol level of 310 mg/dL, only 4% of individuals in the general population would have FH; but this level would identify 95% of those who were first-degree relatives of a patient known to have FH.

Table 3.5 shows values of blood cholesterol required to allow a diagnosis of heterozygous FH with 98% specificity. For a 20-year-old subject, a choles-

TABLE 3.5 — USING TOTAL CHOLESTEROL TO INCREASE THE PROBABILITY OF A CORRECT DIAGNOSIS OF HETEROZYGOUS FAMILIAL HYPERCHOLESTEROLEMIA

Age	Total Cholesterol (mg/dL)		
	1st Degree	2nd Degree	General Population
<18	220	230	270
20	240	250	290
30	270	280	340
40+	290	300	360

Williams RR, et al. *Am J Cardiol.* 1993;72:171-176.

terol of 240 mg/dL or higher suggests the diagnosis in a first-degree relative. On the other hand, without genetic information, a cholesterol reading of 290 mg/dL or greater is needed to suggest such a diagnosis in a general screening. The values required for diagnosis increase with age so that for an individual over 40, a value of 290 mg/dL suggests the diagnosis in a first-degree relative, whereas a person identified only by population screening requires a value of 360 mg/dL or greater. This table assumes great importance in a cost-effective medical environment in which the practitioner does not have the option to pursue DNA studies for the definitive diagnosis.

■ Familial Defective Apolipoprotein B (apo B) Hyperlipoproteinemia

The patient presents with moderate hypercholesterolemia due to a defective apo B. Thus the LDL molecule is not recognized properly by the normally functioning LDL receptors, which are not diminished in number. Unlike FH, there is not a concomitant increase in intermediate-density lipoprotein, so the consequences of the resultant hypercholesterolemia may be less severe. Tendon xanthomas are probably uncommon. An increase in CHD is at least proportional to the increase in LDL-c.

■ Primary Polygenic LDL Excess

This disorder is more prevalent than the specific single-gene disorders. These patients likely represent a heterogeneous group of conditions. Patients do not have xanthomas and less than 10% of first-degree relatives are affected.

Those Disorders Characterized by Mild to Moderate Triglyceride Excess

■ Familial Combined Hyperlipidemia

Familial combined hyperlipidemia (FCHL) is more common than FH, and some researchers estimate it may affect up to 2% of the American population.

Whereas family screening is valuable in FH families, screening of children is not informative because the clinical expression of FCHL is delayed until adulthood. Also, unlike the case in FH, blood cholesterol levels rarely exceed 360 mg/dL and tendon xanthomas are not seen. Both FH and FCHL, however, markedly increase the risk of premature CHD. In the important Seattle study that looked at survivors of MI, FCHL was initially described and found to be the most common inherited cause of premature CHD. Moreover, those with increased apo B despite normal LDL-c values were overrepresented among studies of angiographic CAD. More than a decade later, a placebo-controlled, clinical trial of aggressive therapy examined on angiography individuals with CHD who likely represented cases of FCHL. As an entry criterion, each subject had to have a positive family history of CHD and an elevated apo B of greater than 125 mg/dL. As a group, they proved responsive to aggressive treatment employing either resin and niacin or resin and lovastatin. Those randomized to the treatment groups exhibited, on average, greater decreased progression and increased regression on follow-up coronary angiography. Strikingly, there were significantly fewer clinical events of CHD in the treatment groups as compared with the placebo groups.

This disorder differs from FH in many distinctive ways (Table 3.6). Unlike FH, this genetic disorder is not due to a well-defined, autosomal dominant trait mapped to a single gene locus. FH can be diagnosed confidently in childhood; in FCHL, abnormal lipid levels are expressed later in life. In FH, the dominant lipid abnormality is a high cholesterol, reflecting markedly raised levels of LDL-c. In FCHL, cholesterol levels are not raised as greatly, and high triglycerides and low HDL-c are often noted. Conditions such as hypertension, diabetes, and obesity are common, suggesting a link with hyperinsulinism. Diet and exercise are important modalities of therapy.

■ Familial Dyslipidemic Hypertension

This disorder was described by Williams in his analysis of hypertensive sibships in Utah (Williams et al, 1988). His definition required the diagnosis of essential hypertension before the age of 60 years in two or more siblings who also exhibited one or more abnormalities in lipids. Abnormal lipid values were noted if triglycerides or calculated LDL-c were above the 90th percentile or HDL-c levels were below the 10th percentile. His data estimated that this disorder was seen in 12% of all essential hypertensive individuals and in approximately 25% of such individuals with onset of hypertension before age 60.

■ LDL Phenotype B (Small, Dense LDL)

This apparently atherogenic lipoprotein phenotype was proposed as a genetic marker for risk of CHD

TABLE 3.6 — COMPARISON OF FH AND FCHL		
Features	**FH**	**FCHL**
Childhood hypercholesterolemia	+	–
Tendon xanthomas	+	–
Untreated children and young adults with normal lipid values not likely to develop disorder later in life	+	–
Patients often have high triglycerides/low HDL-c	–	+
Commonly see hypertension, diabetes, obesity in families	–	+
Respond favorably to diet and weight reduction	–	+
Premature CHD in young adults	+	+
Respond well to HMG-CoA reductase inhibitors	+	+
Intravascular ultrasound documenting extensive coronary atherosclerosis, which is angiographically silent	+	+
Aggressive treatment reduces progression of coronary atherosclerosis	+	+

Abbreviations: FH, familial hypercholesterolemia; FCHL, familial combined hyperlipidemia; HDL-c, high-density lipoprotein cholesterol; CHD, coronary heart disease; HMG-CoA, 3-hydroxy-3-methylglutaryl coenzyme A.

Williams RR, et al. *Am J Cardiol*. 1993;72:171-176.

by Austin and co-workers from the Lawrence Berkeley National Laboratory in San Francisco. The authors recruited families from the Mormon community and noted two distinct phenotypes based on gradient gel electrophoretic analysis of LDL subclasses. Phenotype A is characterized by predominance of large buoyant LDL particles; phenotype B, by small, dense

LDL particles. Those with LDL subclass phenotype or pattern B have a trait determined by a major dominant gene which the Berkeley investigators have designated ATHS for atherosclerotic susceptibility; it is linked to a position on the short arm of chromosome 19 and to three other loci. Markers of this trait include elevated triglyceride values and low levels of HDL-c. However, Superko (1996) has noted that simply using triglyceride and HDL-c cutoffs to make this diagnosis can be fraught with error. Furthermore, the LDL pattern B persists even when triglycerides and HDL are normal. Those patients with phenotype or subclass pattern B have several metabolic features which predispose to atherosclerosis. These include elevated levels of small LDL, reduced HDL_2, augmented postprandial lipemia, and insulin resistance. The importance of this trait in the prediction of CHD has been seen in both case-control and prospective cohort studies where those with pattern B have a striking increase in CHD risk. However, as noted in Chapter 2, *Pathophysiology of Hyperlipoproteinemias*, questions remain as to whether increased atherosclerosis is caused by LDL phenotype B per se or by other associated factors.

■ Familial Dysbetalipoproteinemia (a/k/a Familial Type III)

Study of a single large kindred suggested an autosomal dominant mode of inheritance. Most patients with this problem have a structural defect in apo E and present as homozygous cases for the apo E-II isoform. Apo E-II occurs due to a single amino acid substitution (arginine for cysteine) at residue 158. The gene frequency for the allele coding for E-II is about 8%, with homozygosity seen in approximately 1% of the population. The clinical syndrome requires a second metabolic defect or secondary cause. Thus the real frequency of clinical dysbetalipoproteinemia is about 0.02%.

That apo E in humans is secreted predominantly by the liver was dramatically revealed in a case of a man who developed type III hyperlipoproteinemia after receiving a liver transplant. In dysbetalipoproteinemia, remnants of VLDL and chylomicrons accumulate in plasma due to defective binding to liver receptors. Obesity, diabetes, and hypothyroidism are often associated and serve to exaggerate the lipid abnormalities, making detection more obvious. The tendon and tuberous xanthomas have a characteristic appearance. Planar xanthomas in the palmar creases are almost pathognomonic. These individuals also have a propensity for early CHD and peripheral vascular disease. Distinguishing features of familial type III include the distinctive tuboeruptive xanthomas; elevated triglycerides such that cholesterol and triglycerides are often elevated to the same degree; and striking responsiveness to diet, fibrates, or estrogens. Treatment is associated with resolution of the cutaneous xanthomas (Figure 3.6).

■ Familial Hypertriglyceridemia

This is a more benign lipid abnormality. A propensity for early CHD is not seen in these kindreds. Individuals have elevated levels of triglycerides, but cholesterol levels are normal. Apo B levels are normal. The VLDL particles tend to be large and triglyceride rich. Weight control, avoidance of excessive sugars, and exercise are important management strategies. Drug therapy to lower triglycerides is not indicated if there are low-risk levels of LDL-c and HDL-c and there is no associated family history of premature CHD. On the other hand, women who take unopposed estrogen in the face of triglyceride levels elevated above 300 mg/dL do run the risk of a marked increase

FIGURE 3.6 — CUTANEOUS XANTHOMAS IN FAMILIAL TYPE III

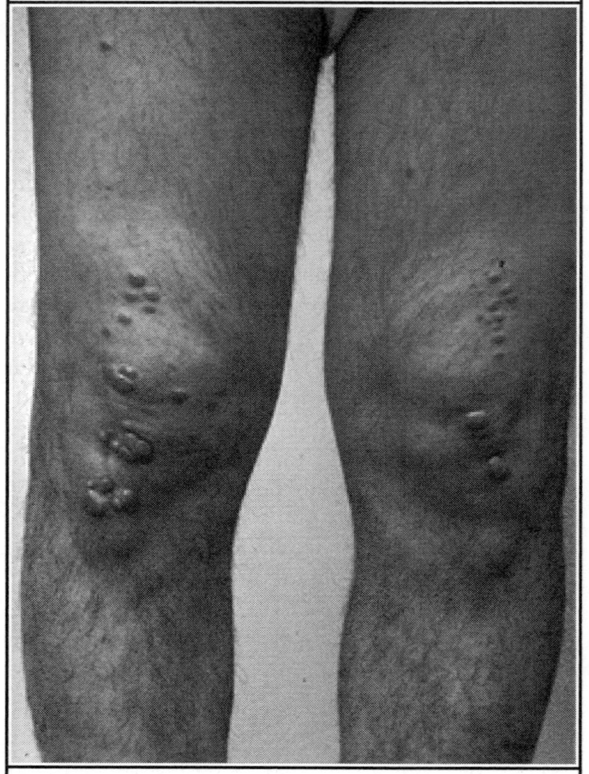

Patient presenting with a distinguishing feature of familial type III: tuboeruptive, cutaneous xanthomas.

in triglycerides (over 1000 mg/dL), which could make them more susceptible to acute pancreatitis.

Those Disorders Characterized by Low Levels of HDL-c or Elevated Values for Lp(a)

■ Familial Hypoalphalipoproteinemia (Low HDL)

Low levels of HDL-c are the hallmark of familial low HDL or familial hypoalphalipoproteinemia. The alpha refers to the position on paper electrophoresis where HDL migrates. The disorder is inherited as an autosomal dominant trait. Affected individuals have values for HDL-c below the 10th percentile as compared with age- and sex-matched controls. In one detailed study of French-Canadian subjects with CAD and very low HDL-c, the HDL was found to be small and dense, perhaps suggesting impaired transport of cholesterol. Nearly all patients with familial syndromes of low HDL are at increased risk of CHD due to their genetic condition (Schaefer, 1984).

■ Excess Lp(a)

Excess Lp(a) is one of the most commonly inherited lipoprotein abnormalities in survivors of MI. Nonetheless, widespread measurement of Lp(a) is not recommended for several reasons. First, not all individuals with high plasma levels of Lp(a) are at equal risk. This may reflect the fact that only a subset of the apo A alleles associated with high plasma levels of Lp(a) may be atherogenic. Second, there is disagreement among the various studies looking at Lp(a) and clinical CHD. Third, there is no generally accepted standardized assay for Lp(a). The differing conditions of storage and differing assays in the above-mentioned trials make conclusions drawn from meta-analyses shaky at best. Fourth, elevated levels of Lp(a) may function, not as a target toward which specific therapy must be directed, but rather as a marker for a high-risk state. In the Familial Atherosclerosis Treatment Study (FATS), although Lp(a) was the best predictor of angiographic severity of CAD, disease regression on the two treatment regimens designed to lower LDL-c correlated with the amount of plasma LDL-c reduction in responders. For the 40 patients with Lp(a) in the upper 10th percentile, events were frequent (39%) if reduction of LDL-c was minimal but were few (9%) if reduction was substantial (Maher et al, 1995). Fifth, other than the use of estrogen and niacin, there are no effective medications to lower Lp(a).

Nonetheless, there are three situations where measurement of Lp(a) may be reasonable. The first two are a personal and family history of premature CHD. Family screening may be particularly useful in these situations. The third case is where the decision to treat is not an easy one. If the Lp(a) is definitely elevated, it may influence one to treat to lower the LDL-c. Data are not available to justify specific Lp(a)-lowering therapy such as niacin, estrogens, or apheresis; rather, the goal is an LDL-c of 100 mg/dL or less.

Disorders Resulting in Severe Hypertriglyceridemia (Chylomicronemia)

Familial hyperchylomicronemia or familial lipoprotein lipase deficiency is a rare genetic disorder. It is characterized by fasting plasma showing marked triglyceride excess of 1000 mg/dL or more and chylomicronemia visible to the naked eye as a creamy supernatant when a tube of fasting plasma is kept refrigerated (Figure 3.1). It is inherited as an autosomal recessive trait. Deficiency or absence of either the major enzyme required for metabolism of triglyceride-laden chylomicrons and VLDL, called lipoprotein lipase, or the apoprotein which serves as a necessary cofactor for lipoprotein lipase action, known as apo C-II, define this inherited disorder (Table 3.7).

TABLE 3.7 — EFFECT OF TREATMENT ON LIPIDS IN SCOR FH INTERVENTION TRIAL

	Control Group		Treatment Group	
	BL	**OT**	**BL**	**OT**
Total Cholesterol	367	335	378	261
HDL-c	51	51	47	59
LDL-c	274	243	283	172
Decline in LDL-c (%)	—	11.3	—	39.2
Increase in HDL-c (%)	—	0	—	25.5

Abbreviations: SCOR, Specialized Center for Atherosclerosis Research; FH, familial hypercholesterolemia; BL, baseline; OT, on trial; HDL-c, high-density lipoprotein cholesterol; LDL-c, low-density lipoprotein cholesterol.

Adapted from Kane JP, et al. *JAMA*. 1990;264:3007-3012.

The most serious symptom is recurrent abdominal pain, often beginning in childhood and heralding an attack of acute pancreatitis. Serum and urine amylase levels are often not elevated, due either to interference by the excessive plasma lipids with the amylase assay and/or to circulating inhibitors. The diagnosis of acute pancreatitis must be suspected on clinical grounds and confirmed with an imaging study, such as ultrasound or computed tomography scan. During periods of severe hypertriglyceridemia, the patient presents with a characteristic clinical picture of lipemia retinalis, eruptive xanthomas, and hepatosplenomegaly. The xanthomas are yellow-orange papules on a reddish base, reflecting macrocytes engorged with chylomicrons (Figure 3.7). They are often

FIGURE 3.7 — ERUPTIVE XANTHOMAS IN SEVERE HYPERTRIGLYCERIDEMIA

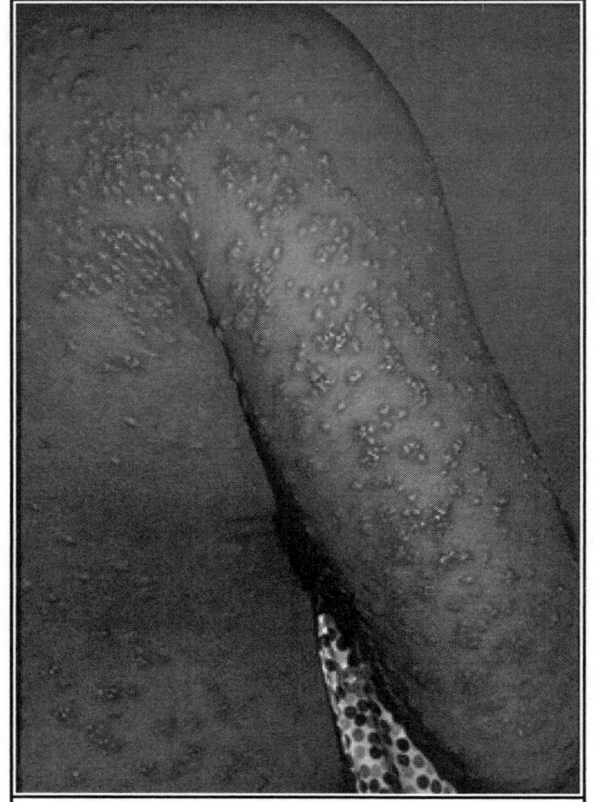

Eruptive xanthomas are yellow-orange papules on a reddish base, reflecting macrocytes engorged with chylomicrons.

missed because of their location and recede rapidly with effective triglyceride lowering.

Patients with total absence of lipoprotein lipase activity do not appear to be predisposed to early coronary events.

SUGGESTED READINGS

Abate N, Vega GL, Grundy SM. Variability in cholesterol content and physical properties of lipoproteins containing apolipoprotein B-100. *Atherosclerosis*. 1993;104:159-171.

Brown WV, Breslow JL. Genetic lipoprotein abnormalities producing high blood cholesterol. *Heart Dis Stroke*. 1992;1:405-407.

Fredrickson DS. Dyslipoproteinemia from phenotypes to genotypes... a remarkable quarter century. *Circulation*. 1993;87(suppl 3):S1-S59.

Genest J Jr, McNamara JR, Ordovas JM, et al. Lipoprotein cholesterol, apolipoprotein A-I and B and lipoprotein (a) abnormalities in men with premature coronary artery disease. *J Am Coll Cardiol*. 1992;19:792-802.

Goldstein JL, Brown MS. Familial hypercholesterolemia. In: Scriver CR, Beaudet AL, Sly WF, Valle D, eds. *Metabolic Basis of Inherited Disease*. 6th ed. New York, NY: McGraw-Hill Book Co; 1990:1215-1250.

Hausmann D, Johnson JA, Sudhir K, et al. Angiographically silent atherosclerosis detected by intravascular ultrasound in patients with familial hypercholesterolemia and familial combined hyperlipidemia: correlation with high-density lipoproteins. *J Am Coll Cardiol*. 1996;27:1562-1570.

Hirany S, Li D, Jialal I. A more valid measurement of low-density lipoprotein cholesterol in diabetic patients. *Am J Med*. 1997;102:48-53.

Innerarity TL, Mahley RW, Weisgraber KH, et al. Familial defective apolipoprotein B-100: a mutation of apolipoprotein B that causes hypercholesterolemia. *J Lipid Res*. 1990;31:1337-1349.

Kane JP, Malloy MJ, Ports TA, Phillips NR, Diehl JC, Havel RJ. Regression of coronary atherosclerosis during treatment of familial hypercholesterolemia with combined drug regimens. *JAMA*. 1990;264:3007-3012.

Maher VM, Brown BG, Marcovina SM, Hillger LA, Zhao XQ, Albers JJ. Effects of lowering elevated LDL cholesterol on the cardiovascular risk of lipoprotein(a). *JAMA*. 1995;274:1771-1774.

Mahley RW, Weisgraber KH, Innerarity TL, Rall SC Jr. Genetic defects in lipoprotein metabolism. Evaluation of atherogenic lipoproteins caused by impaired catabolism. *JAMA*. 1991;265:78-83.

Marcil M, Boucher B, Krimbou L, et al. Severe familial HDL deficiency in French-Canadian kindreds. Clinical, biochemical and molecular characterization. *Arterioscler Thromb Vasc Biol*. 1995; 15:1015-1024.

Santamarina-Fojo S, Brewer HB Jr. The familial hyperchylomicronemia syndrome. New insights into underlying genetic defects. *JAMA*. 1991;265:904-908.

Schaefer EJ. Clinical, biochemical, and genetic features in familial disorders of high density lipoprotein deficiency. *Arteriosclerosis*. 1984;4:303-322.

Schaefer EJ, Gregg RE, Ghiselli G, et al. Familial apolipoprotein E deficiency. *J Clin Invest*. 1986;78:1206-1219.

Serfaty-Lacrosniere C, Civeira F, Lanzberg A, et al. Homozygous Tangier disease and cardiovascular disease. *Atherosclerosis*. 1994;107:85-98.

Superko HR. Lipid disorders contributing to coronary heart disease: an update. *Curr Probl Cardiol*. 1996;21:736-780.

Williams RR, Hopkins PN, Hunt SC, et al. Population-based frequency of dyslipidemia syndromes in coronary-prone families in Utah. *Arch Intern Med*. 1990;150:582-588.

Williams RR, Hunt SC, Hopkins PN, et al. Familial dyslipidemic hypertension. Evidence from 58 Utah families for a syndrome present in approximately 12% of patients with essential hypertension. *JAMA*. 1988;259:3579-3586.

Williams RR, Hunt SC, Schumacher C, et al. Diagnosing heterozygous familial hypercholesterolemia using new practical criteria validated by molecular genetics. *Am J Cardiol*. 1993;72:171-176.

Zulewski H, Miserez AR, Sieber CC, Chiodetti N, Keller U. Acquired type III hyperlipoproteinemia in recipient of liver transplant. *Lancet*. 1994;343:971.

4 Causes of Secondary Hyperlipoproteinemia

Secondary or acquired causes of hyperlipidemia are important for several reasons:

- First, they emphasize that lipid abnormalities may be clues to uncovering underlying disorders which require specific treatment.
- Second, they may explain why a previously treatable genetic lipid disorder has become seemingly resistant to treatment.
- Third, they may point to an alternate, potentially safer form of therapy.

The mnemonic of 4 Ds is a useful way to view the secondary causes of hyperlipidemia:

- **D**iet
- **D**rugs
- **D**isorders of metabolism
- **D**iseases.

Table 4.1 shows how common secondary causes can be listed as a function of the predominant lipid/lipoprotein abnormality. Thus, depending on the lipoprotein(s) which are abnormal, the clinician can go to this chart to determine at a glance which secondary causes might be operative.

Diet

Diet itself is covered in detail in Chapter 8, *Dietary Therapy for Hyperlipidemia.* The clinician would be wise to inquire about dietary saturated fat and dietary cholesterol when approaching a patient

TABLE 4.1 — OVERVIEW OF ACQUIRED CAUSES OF DYSLIPIDEMIA

Secondary Cause	High Cholesterol (High LDL-c)	Triglyceride Excess Mild to Moderate (High VLDL)	Low HDL-c	Severe Triglyceride Excess: Chylomicronemia Syndrome
Dietary	Saturated fats, caloric excess, anorexia	Weight gain, alcohol	Low-fat diet, sweetened foods, sugar	Alcohol and fat plus genetic lipid disorder
Drugs	Diuretics, cyclosporine, glucocorticoids, rosiglitazone, fibrates, fish oil	Retinoic acid, beta adrenergic blockers, estrogens, glucocorticoids, protease inhibitors	Anabolic steroids, most progestins, beta-blockers, cigarettes	Glucocorticoids, estrogens plus genetic lipid disorder
Disorders of metabolism	Hypothyroidism, pregnancy, DM	Obesity, type 2 DM, pregnancy (third trimester)	Obesity, type 2 DM	Diabetes, hypothyroidism plus genetic lipid disorder
Diseases	Nephrotic syndrome, biliary obstruction	Chronic renal failure ± dialysis, nephrotic syndrome	Chronic renal failure ± dialysis	Systemic lupus erythematosus, lymphoma (rare)

Abbreviations: LDL-c, low-density lipoprotein cholesterol; VLDL, very low-density lipoprotein; HDL-c, high-density lipoprotein cholesterol; DM, diabetes mellitus.

Adapted from Stone NJ. *Med Clin North Am*. 1994;78:120.

concerning changes in cholesterol and low-density lipoprotein cholesterol (LDL-c); weight change, alcohol intake, and carbohydrate intake should be questioned when confronted with lipid changes involving triglycerides and/or high-density lipoprotein cholesterol (HDL-c).

One point worthy of mention is that anorexia nervosa can occasionally present with striking hypercholesterolemia; this occurs in varying degrees. Sometimes it merely indicates a person has become overly compulsive about a low-fat diet. A putative mechanism is the diminished cholesterol and bile acid turnover secondary to a reduced caloric intake. The therapy seems almost counterintuitive to the patient because more food is prescribed, not less. Yet, as the patient increases the intake of a more nutritionally balanced diet, the cholesterol falls.

Drugs

Drug therapy must be considered as an aggravating factor to the condition of any patient under clinical scrutiny for tight lipid control whose lipid profile suddenly worsens. Several general rules are worth reviewing in this regard:

- First, it is important to view any medication that the patient is receiving as having the potential to increase either the risk of coronary heart disease or pancreatitis. Although the focus is on lipids and coronary risk, the concern regarding pancreatitis is not trivial. The addition of oral steroids or estrogen in a person with an underlying genetic disorder involving triglyceride metabolism (and particularly if there are other acquired causes of triglyceride excess, such as obesity or increased alcohol intake) can have catastrophic results as triglycerides may rise precipitously, causing acute pancreatitis.

- Second, the risks and benefits of substituting another medication must be viewed in the context of the entire clinical picture. Although beta blockers raise triglycerides and lower HDL-c, the important effect of beta-blockers of prolonging life after an acute myocardial infarction (MI) must be considered. The wiser strategy is to continue the beta-blockers, which have been shown to be beneficial for their antiarrhythmic, anti-ischemic, and antihypertensive properties, and use diet, exercise in a cardiac rehabilitation setting, and drug therapy to control the dyslipidemia (Table 4.1).

A comprehensive review of the effects of drugs on lipids by Henkin and co-workers (1992) is a valuable resource for physicians who desire information beyond what can be presented in this chapter. This book concentrates on the drugs that are used most often; the discussion is divided into cardiovascular drugs, endocrine drugs, and others.

■ Cardiovascular Drugs

Diuretics are important drugs in the treatment of hypertension as they have been shown to prolong life and reduce cardiovascular risk. Yet in short-term studies, diuretics raise cholesterol and LDL-c as well as triglycerides. Moser (1989) observed that significant elevations in cholesterol were not seen at 1 year in most hypertension treatment trials. More recently, there has been an appreciation that patients who stop their medications in these trials have a significant impact on the conclusions due to "intention to treat" analysis employed. In a trial of 634 consecutive hypertensive patients treated at the University of Chicago, the changes seen over time were much greater in the "on-therapy" group as contrasted with the "intention-to-treat" group (Figure 4.1).

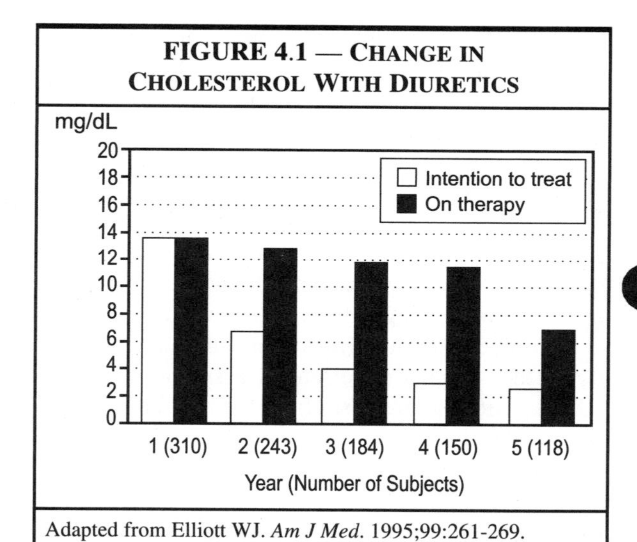

FIGURE 4.1 — CHANGE IN CHOLESTEROL WITH DIURETICS

Adapted from Elliott WJ. *Am J Med.* 1995;99:261-269.

There is a dose-response effect with thiazides, and small doses are not likely to make a great difference. Actually, the small increase in lipids reported is potentially of more concern in younger hypertensives than in the elderly, who are sensitive to small amounts of diuretics. In men and in postmenopausal women, high doses of chlorthalidone (100 mg/d) do appear to raise LDL-c.

Beta-adrenergic blocking agents are also highly effective drugs used for the treatment of hypertension, arrhythmias, and ischemic heart disease. In the Beta Blocker Heart Attack Trial (BHAT), which looked at men post-MI, beta-blocker usage was associated with a 17% higher triglyceride level and a 6% lower HDL-c level with no significant effect on LDL-c. As noted above, despite these effects, beta-blockers should not be stopped or substituted for in patients post-MI; diet, exercise, and lipid-lowering drugs should be used in complementary fashion.

Stage I diastolic hypertension treatments were examined in detail in the Treatment of Mild Hypertension Study (TOMHS), using treatment-by-gender groupings. These results must be put in perspective because there were also changes in weight and exercise, which significantly affect lipid levels. In this study, men lost significantly more weight (4.1 kg vs 2.4 kg) and increased their physical activity to a greater extent than women. With the exception of the men who were assigned to treatment with acebutolol (a beta-blocker with intrinsic sympathomimetic activity), other groups experienced a small increase in HDL-c, with the greatest effect seen in the group receiving the alpha-blocker doxazosin. Among women, total and LDL-c levels increased in the chlorthalidone-treated group, but these did not differ significantly from the placebo group. The conclusion is that usual treatments for hypertension when accompanied by weight loss and an increase in exercise do not have clinically adverse effects on the lipid profile.

Amiodarone is a cardiovascular drug. It can elevate cholesterol independently of its effects on the thyroid hormone. Thyroid abnormalities are commonly seen due to the high iodine content of this drug, so elevations of cholesterol must also make the clinician suspect thyroid hypofunction.

■ Endocrine Drugs

Anabolic steroids cause a marked decrease in HDL-c and especially the HDL_2 fraction. This decrease can occur within 1 to 2 weeks of beginning therapy. Since this situation greatly augments cardiac risk, the use of these drugs should be actively discouraged. A comparison of testosterone with stanozolol showed that while stanozolol causes a 29% increase in LDL-c and decreases HDL-c by 33%, parenteral testosterone caused a decrease of 16% in LDL-c and only a 9% decrease in HDL-c. Thus parenteral testosterone did not adversely affect the LDL-c/HDL-c ratio.

Glucocorticoids have potent effects on lipid levels. The clinician should know the underlying lipid status of the patient before starting a course of either prolonged or high-dose steroid therapy. Prednisone in healthy men increases very low-density lipoprotein (VLDL) production and stimulates formation of HDL. High-dose steroid therapy applied to a patient with significant hypertriglyceridemia, as is seen in diabetes not under good control, can cause severe lipemia. The increased weight gain and potential glucose intolerance with steroid therapy further compounds the effects of steroids on lipid metabolism.

Estrogens and androgens have, for the most part, opposing effects on lipid metabolism. Many progestational agents have some androgenic activity. Estrogens raise triglycerides by increasing VLDL production and also raise HDL-c; LDL-c is lowered. These changes were well-documented in the Postmenopausal Estrogen/Progestin Intervention (PEPI) trial (Table 4.2). Unopposed estrogen, which is suitable for the postmenopausal woman without a uterus, clearly has the most advantageous lipid effects. One caution regarding unopposed estrogen therapy is the use of such therapy in a woman with an underlying lipid disorder. Estrogen therapy is a common cause of the chylomicronemia syndrome in women with an underlying genetic lipid disorder. Thus baseline lipid values should be obtained in women before starting hormone replacement therapy. In those patients with triglyceride values above 300 mg/dL, programs of diet and exercise to reduce the elevated triglyceride values may allow reductions in baseline triglycerides and reduce the risk of an excessive rise in triglyceride values. For those women who cannot take oral estrogens due to this untoward effect, transdermal estro-

TABLE 4.2 — EFFECTS OF ESTROGENS AND/OR PROGESTINS ON BLOOD LIPIDS: THE PEPI TRIAL			
Preparation	**LDL-c**	**Triglycerides**	**HDL-c**
Premarin 0.625	– 14.5	13.7	5.6
Premarin, micronized progesterone (cycle)	– 14.8	13.4	4.3
Premarin, Provera 10 (cycle)	– 17.7	12.7	1.6
Premarin, Provera 2.5 (continuous)	– 16.5	11.4	1.2
Placebo	– 4.1	– 3.2	– 1.2

Abbreviations: PEPI, Postmenopausal Estrogen/Progestin Intervention; LDL-c, low-density lipoprotein cholesterol; HDL-c, high-density lipoprotein cholesterol.

Adapted from The Writing Group for the PEPI Trial. *JAMA*. 1995;273:199-208.

gen has a minimal effect on triglycerides in contrast to those changes seen when oral estrogens are given.

Progestational agents vary in their ability to affect lipid levels. Although many progestational agents have some androgenic activity, some newer ones actually have estrogenlike effects on lipoproteins; these are desogestrel, norgestimate, and gestodene. Progestins have negative effects with regard to vessel-wall reactivity and should be used in the lowest possible effective doses. Oral contraceptive formulations have changed over the years. Effects on HDL-c depend on the relative proportions of estrogen and progesterone as well as on the androgenicity of the progesterone. Again, it may be of particular value to check lipid levels in sedentary, overweight women before oral contraceptives are prescribed. Exercise programs can help improve triglyceride levels in such patients.

Tamoxifen in normolipidemic women is associated with a reduction in LDL-c and Lp(a) without major effects on triglycerides or HDL-c. In women with genetic hypertriglyceridemia, however, tamoxifen and a compound with a similar structure, clomiphene, can markedly increase triglyceride levels and cause pancreatitis (Glueck et al, 1994; Castro et al, 1999). Raloxifene, a selective estrogen-receptor modulator that does not stimulate the endometrium of postmenopausal women, lowers total cholesterol and LDL-c but, unlike natural estrogen, does not raise HDL-c (Delmas et al, 1997).

■ Other Frequently Used Medications Affecting Lipids

Retinoids (isotretinoin) are used for acne but can cause mild to marked elevations in triglycerides with a decrease in effect on HDL-c. Peak levels of triglycerides can occur as early as 4 weeks in men and 12 weeks in women. These lipid changes cease promptly when the medication is discontinued. Lipid levels should, therefore, be checked before therapy is begun. For those patients who are sedentary and overweight, attention to diet and exercise before beginning such therapy can be expected to ameliorate the changes seen. Prudent use suggests that retinoids not be given to those with triglyceride elevations at baseline and that lipid levels be followed monthly for the first several months to determine the individual response.

Cyclosporine is a mainstay for transplantation programs. It is highly lipid-soluble, and a significant portion is bound to lipoproteins. It increases LDL substantially along with Lp(a). In addition, severe hypertriglyceridemia has been described after bone marrow transplantation. The most important caution concerning its use is the interaction with HMG-CoA reductase inhibitors. Although these drugs can be used safely with cyclosporine, drug dosages are kept lower than might commonly be used to avoid precipitating myositis, which can lead to rhabdomyolysis. Most centers do not routinely use more than the equivalent of 20 mg of lovastatin or pravastatin in these patients who must be aware of this potential toxic interaction.

The immunosuppressant medication sirolimus can increase both cholesterol and triglyceride levels. Data on the effects of this drug on specific lipoprotein classes are not available at the time of this writing.

Protease inhibitors have improved the prognosis of those with advanced human immunodeficiency virus infection. Use of these drugs, particularly ritonavir, can lead to lipid complications (Sullivan and Nelson, 1997; Henry et al, 1998). The characteristic syndrome includes central adiposity with wasting of the face and extremities, dyslipidemia with elevated triglycerides and low HDL-c, and insulin resistance. The mean onset of these effects is 6 to 10 months after the start of drug treatment. It is postulated that protease inhibitors inhibit proteins involved in lipid metabolism.

■ HDL and Drugs

Beta-blocker eye drops may adversely affect lipid profiles. Topical 0.5% timolol eye drops can reduce HDL 9% and increase triglycerides 12%. Thus, inquiry about eye medication should be part of the appraisal of any patient with low HDL-c. Cigarettes are much more efficient nicotine delivery systems than cigars or pipes; therefore, cigarette smokers can have reduced levels of HDL-c while cigar and pipe smokers do not.

On the other hand, cimetidine may increase HDL in women but not in men, whereas ranitidine does not have this effect. *Phenobarbital*, *phenytoin*, and *carbamazepine* often appear to be associated with substantial increases in HDL-c. Beta-stimulating drugs such as terbutaline increase HDL-c. This may explain the increased HDL-c seen in those with chronic obstructive airway disease. Due to the absence of clinical data and since these drugs may cause important side effects, it is not considered routine practice to use them to raise HDL-c in those with lowered levels of HDL-c.

Disorders of Metabolism

■ Hypothyroidism

Hypothyroidism is one of the most common causes of a raised total cholesterol and LDL-c level. It should be suspected clinically in patients with previously lower cholesterol values who now present with raised cholesterol values. In one study, there was a prevalence of 4.4% when individuals with values over 310 mg/dL had T_4 and thyroid stimulating hormone (TSH) checked. An additional 9% were found with raised TSH levels only, suggesting subclinical hypothyroidism. Hypothyroidism can aggravate lipid levels in patients with familial hyperlipidemia to the degree that they come to clinical attention. It is good clinical practice to check a TSH level when a patient with genetic hyperlipidemia that was previously well controlled on a diet and/or drug regimen suddenly presents with an elevated LDL-c which cannot be easily explained by altered compliance with a previously effective regimen.

■ Obesity

Markedly obese subjects may have normal lipids. Those with abdominal obesity with increased waist-to-hip ratios (who look like "apples" instead of "pears") often have accompanying increased triglycerides and reduced HDL-c along with other signs of hyperinsulinemia, such as hypertension and glucose intolerance. (This subject is covered more comprehensively in Chapter 11, *Special Populations*.)

■ Diabetes Mellitus

Diabetics can exhibit the entire range of lipid disorders as the lack of insulin affects both cholesterol and triglyceride metabolism. Poor control due to insulin deficiency coupled with increased fat intake to cover increased urinary losses can lead to diabetic lipemia with all of the characteristics of the chylomicronemia syndrome. Diabetic treatment can affect lipid profiles as well. Insulin therapy may lower LDL-c, while insulin sensitizers, such as rosiglitazone and pioglitazone, improve glycemic profiles but may raise total cholesterol and LDL-c (Fonseca et al, 2000). (This subject is covered more comprehensively in Chapter 11, *Special Populations*.)

■ Pregnancy and Lactation

With each trimester, cholesterol and triglyceride concentrations rise. The major increase in cholesterol is seen in the second trimester. For most women, this occurrence is not of clinical importance. Women with genetic lipid disorders and raised levels of triglycerides at the onset of pregnancy are at risk for a marked acceleration of the triglyceride values in the third trimester. When triglyceride values exceed 2000 mg/dL, these women are at risk for the chylomicronemia syndrome and pancreatitis. The best treatment is prevention with regular exercise and weight control advised before pregnancy. If triglycerides remain strictly controlled, these patients can go through pregnancy without serious complication. On the other hand, for those who present in the third trimester with triglycerides greater than 2000 mg/dL, hospitalization and strict control of fat intake, insulin (if they are overtly diabetic), and careful assessment are required. After delivery, cholesterol and triglyceride values fall significantly; and by 6 weeks postpartum, triglyceride values are usually at baseline. Women whose lipid levels during pregnancy exceed the 95th percentile often develop overt hyperlipidemia with aging.

Nonmetabolic Diseases

■ Liver Disease

Cholestasis can cause marked hypercholesterolemia. Patients with primary biliary cirrhosis present with cholesterol levels that can exceed 500 mg/dL. They have cutaneous xanthomas, but eventually the origin of their problem is signified by the accompanying pruritus and jaundice. An abnormal lipoprotein known as lipoprotein-X (Lp-X) is present along with elevated levels of LDL. Lp-X is actually a complex molecule of biliary lecithin, free cholesterol, albumin, and apo C. Usual modes of therapy are not effective.

In those patients with hepatocellular damage, hepatic production of apolipoproteins and enzymes may be impaired. Niacin can cause hepatotoxicity, and often the hallmark is very low levels of cholesterol and LDL-c. Accordingly, when values too good to believe are seen with any therapy that can cause hepatotoxicity, liver function tests should be checked. If liver function tests are abnormal, indicating liver toxicity from medication, other potential hepatotoxic substances, such as alcohol or other drugs, should be stopped as well to allow the liver to recover.

■ Renal Disease

Patients with chronic renal failure have increased triglycerides and low HDL-c with enrichment of HDL and LDL with triglycerides. Increasing the efficiency of dialysis only partially corrects the abnormality. Peritoneal dialysis may cause higher, more sustained levels of triglycerides, owing to resorption of large amounts of glucose from peritoneal fluid which stimulate VLDL production from the liver. Patients with chronic renal failure should be considered to have an atherogenic profile with chylomicron and VLDL remnants, intermediate-density lipoproteins, and small, dense LDL.

Nephrotic syndrome is characterized by proteinuria, hypoalbuminemia, hypercholesterolemia, and, later, edema. The presenting picture can be pure hypercholesterolemia or a combined hyperlipidemia. Usually, in cases with lower albumin levels, triglycerides are elevated as well. Grundy and co-workers found that patients with nephrosis who presented with hypercholesterolemia, when carefully studied, generally had the following: (Vega et al, 1995)

- Lower fractional catabolic rates of LDL apo B than normolipidemic healthy individuals
- Cholesterol-rich LDL particles
- No overproduction of LDL apo B.

In contrast, patients with combined hyperlipidemia were found to have these characteristics:

- High fractional catabolic rates for LDL apo B compared with normolipidemic controls
- Cholesterol-poor LDL particles
- Markedly elevated production rates for LDL.

Also, for the group as a whole, there was a positive correlation between plasma triglyceride levels and fractional catabolic rates. Thus the metabolism of LDL is strikingly different between the two forms of nephrotic dyslipidemia. Lp(a) is elevated in nephrosis and improves when angiotensin-converting enzyme inhibitors are used.

Nephrotic syndrome should be suspected in the patient whose condition has developed a refractory hypercholesterolemia. A recent consultation involved a woman with familial hypercholesterolemia whose condition suddenly could not be controlled with her usual drug therapy. This prompted a consideration of secondary causes. A urine sample for protein confirmed the diagnosis, which, more importantly, led to corrective treatment. As expected, the patient's response to treatment of her nephrotic syndrome was paralleled by improvement in her cholesterol values.

Transplantation

The dyslipidemia of transplant patients may have multiple causes:

- Weight gain
- Cumulative use of steroids
- Concomitant drug therapy, including cyclosporine and sirolimus
- Development of diabetes.

The Mayo Clinic experience underscored the importance of weight control. The nondiabetic, hypercholesterolemic group had significantly greater mean values for weight gained (11.2 kg) than those with lower values for cholesterol. In this study, prednisone and serum creatinine levels were similar in both groups. Reducing prednisone dosage or going to alternate-day therapy, if possible, may improve the triglyceride levels in adults but not necessarily in children. Studies of cardiac transplantation have shown marked increases in LDL-c, which seems best explained by prednisone therapy, age, and preoperative cholesterol level. Lp(a) levels are reported to decrease after transplantation, perhaps due in part to effects of the immunosuppression therapy.

Hyperlipidemia is a significant problem after heart transplantation. There may be an association between hyperlipidemia and development of accelerated vascular disease (Ballantyne et al, 1997). Lipid-lowering therapy is important in the long-term management of these patients (Stapleton et al, 1997).

SUGGESTED READINGS

Ballantyne CM, el Masri B, Morrisett JD, Torre-Amione G. Pathophysiology and treatment of lipid perturbation after cardiac transplantation. *Curr Opin Cardiol.* 1997;12:153-160.

Bershad S, Rubinstein A, Paterniti JR, et al. Changes in plasma lipids and lipoproteins during isotretinoin therapy for acne. *N Engl J Med.* 1985;313:981-985.

Byington RP, Worthy J, Craven T, Furberg CD. Propranolol-induced lipid changes and their prognostic significance after a myocardial infarction: the Beta-Blocker Heart Attack Trial Experience. *Am J Cardiol.* 1990;65:1287-1291.

Castro MR, Nguyen TT, O'Brien T. Clomiphene-induced severe hypertriglyceridemia and pancreatitis. *Mayo Clinic Proc.* 1999;74: 1125-1128.

Coleman AL, Diehl DL, Jampel HD, Bachorik PS, Quigley HA. Topical timolol decreases plasma high-density lipoprotein cholesterol level. *Arch Ophthalmol.* 1990;108:1260-1263.

Delmas PD, Bjarnason NH, Mitlak BH, et al. Effects of raloxifene on bone mineral density, serum cholesterol concentrations, and uterine endometrium in postmenopausal women. *N Engl J Med.* 1997;337:1641-1647.

Elliott WJ. Glucose and cholesterol elevations during thiazide therapy: intention-to-treat versus actual on-therapy experience. *Am J Med.* 1995;99:261-269.

Fonseca V, Rosenstock J, Patwardhan R, Salzman A. Effect of metformin and rosiglitazone combination therapy in patients with type 2 diabetes mellitus: a randomized controlled trial. *JAMA.* 2000;283:1695-1702.

Glueck CJ, Lang J, Hamer T, Tracy T. Severe hypertriglyceridemia and pancreatitis when estrogen replacement therapy is given to hypertriglyceridemic women. *J Lab Clin Med*. 1994;123:59-64.

Gonyea JE, Anderson CF. Weight change and serum lipoproteins in recipients of renal allografts. *Mayo Clin Proc*. 1992;67:653-657.

Henkin Y, Como JA, Oberman A. Secondary dyslipidemia. Inadvertent effects of drugs in clinical practice. *JAMA*. 1992;267:961-968.

Henry K, Melroe H, Huebsch J, et al. Severe premature coronary artery disease with protease inhibitors. *Lancet*. 1998;351:1328. Letter.

Keilani T, Schlueter WA, Levin ML, Batlle DC. Improvement of lipid abnormalities associated with proteinuria using fosinopril, an angiotensin-converting enzyme inhibitor. *Ann Intern Med*. 1993;118:246-254.

Knopp RH. Cardiovascular disease in women: relevance to clinical lipidology. *Lipid Letter*. 1996;2:1-3.

Lewis CE, Grandits A, Flack J, McDonald R, Elmer PJ. Efficacy and tolerance of antihypertensive treatment in men and women with stage 1 diastolic hypertension. Results of the Treatment of Mild Hypertension Study. *Arch Intern Med*. 1996;156:377-385.

Moser M. Lipid abnormalities and diuretics. *Am Fam Physician*. 1989;40:213-220.

Series JJ, Biggart EM, O'Reilly DS, Packard CJ, Shepherd J. Thyroid dysfunction and hypercholesterolemia in the general population of Glasgow, Scotland. *Clin Chim Acta*. 1988;172:217-221.

Stapleton DD, Mehra MR, Dumas D, et al. Lipid-lowering therapy and long-term survival in heart transplantation. *Am J Cardiol*. 1997;80:802-805.

Stone NJ. Secondary causes of hyperlipidemia. *Med Clin North Am*. 1994;78:117-141.

Sullivan AK, Nelson MR. Marked hyperlipidaemia on ritonavir therapy. *AIDS*. 1997;11:938-939. Letter.

The Writing Group for the PEPI Trial. Effects of estrogen or estrogen/progestin regimens on heart disease risk factors in postmenopausal women. The Postmenopausal Estrogen/Progestin Interventions (PEPI) Trial [published erratum appears in *JAMA*. 1995;274: 1676]. *JAMA*. 1995;273:199-208.

Thompson PD, Cullinane EM, Sady SP, et al. Contrasting effects of testosterone and stanozolol on serum lipoprotein levels. *JAMA*. 1989;261:1165-1168.

Vega GL, Toto RD, Grundy SM. Metabolism of low density lipoproteins in nephrotic dyslipidemia: comparison of hypercholesterolemia alone and combined hyperlipidemia. *Kidney Int*. 1995;47: 579-586.

5 Clinical Appraisal and Goals of Therapy

Defining Risk

Clinical screening consists of an evaluation of the cholesterol level in the context of the total coronary heart disease (CHD) risk-factor profile (Table 5.1). An elevated cholesterol level in the absence of any other risks for atherosclerosis will not have the same significance in predicting a coronary event as it will in the subject who has several other risk factors. In addition, the effect of the various risk factors is often more than additive; that is, the magnitude of two or more risk factors together may be more than the sum of the magnitude of each of them. Eliciting the total coronary risk profile is also important because it can then allow intervention of several factors that may help decrease overall risk even more than a single risk. Finally, those with established atherosclerosis have the highest risk of a near-term coronary event. Defining the cholesterol profile and risk-factor status is of the highest priority in these patients.

The evaluation of overall risk begins in the clinical encounter with eliciting a history of the various factors known to increase risk of coronary events. The CHD risk factors have been identified and characterized over the past 50 years in a variety of epidemiologic and clinical studies. The major risk factors, other than cholesterol, are listed in Table 5.1. The National Cholesterol Education Program (NCEP) guidelines recognize that low high-density lipoprotein cholesterol (HDL-c) levels markedly increase the risk of CHD and very high HDL-c levels are associated with longev-

TABLE 5.1 — RISK FACTORS TO CONSIDER WHEN CONTEMPLATING LIPID-LOWERING THERAPY
Nonmodifiable
• *Age and Gender*—Male ≥45, female ≥55 (or premature menopause without estrogen replacement)
• *Family History*—Definite myocardial infarction or sudden death before age 55 in male first-degree relative; before age 65 in female first-degree relative
Modifiable
• Cigarette smoking
• Hypertension
• Low high-density lipoprotein cholesterol (<35 mg/dL); HDL-c ≥60 mg/dL is a negative risk factor
• Diabetes mellitus
Other Intervention Targets
• Sedentary lifestyle
• Obesity

ity. Thus an HDL-c level below 35 mg/dL will increase the tally of risk factors by one; an HDL-c level ≥60 mg/dL will reduce the tally of risk factors by one. In addition, the NCEP guidelines do not specifically identify risk factors such as sedentary lifestyle and obesity (see Chapters 9, *Exercise and Lipids,* and 11, *Special Populations*). These factors are, however, specified as targets for intervention in coronary risk reduction. In fact, several studies have demonstrated the potential benefit of addressing these risk factors (Tables 5.2 and 5.3).

The NCEP suggests that total blood cholesterol and HDL-c be evaluated in all people at age 20 and at approximately 5-year intervals thereafter. A full lipid profile need not be done as an initial screen. Fasting is only necessary if triglycerides are to be measured and low-density lipoprotein cholesterol (LDL-c) calculated. Actions to be taken, according to these guidelines, include the following:

TABLE 5.2 — RISK FACTORS (MAGNITUDE OF EFFECT OF SOME FACTORS)	
Risk Factor	**Odds Ratio**
Smoker (current)	3.6
Smoker (≥10 cigarettes per day)	6.7
History of hypertension	2.69
Diabetes mellitus	2.64
Waist-to-hip ratio (central obesity)	1.62 per SD increase
Smoking and elevated glucose	10.6
Vegetarianism	0.55
SES group (highest vs lowest)	0.32
Abbreviations: SD, standard deviation; SES, socioeconomic status.	
Pais P, et al. *Lancet*. 1996;348:358-363.	

TABLE 5.3 — RISK-FACTOR INTERVENTION FOR REDUCTION OF MYOCARDIAL INFARCTION	
Lifestyle Change	**Possible ↓ MI Risk**
Smoking cessation	50% to 70%
Exercise	45%
Hypertension control	2% to 3% for each mm reduction in diastolic BP
Postmenopausal estrogen replacement	44%
Aspirin	33%
Abbreviations: MI, myocardial infarction; BP, blood pressure.	
Adapted from Manson JE, et al. *N Engl J Med*. 1992;326:1406-1416.	

- If the initial total cholesterol is <200 mg/dL, in the absence of other risks, no further action need be taken.
- If the cholesterol is between 200 and 230 mg/dL, with an HDL-c over 35 mg/dL and fewer than two other risk factors, the only action that should be taken is to provide information on diet and exercise (see Chapters 8, *Dietary Therapy for Hyperlipidemia,* and 9, *Exercise and Lipids*) and to consider repeat testing in 1 to 2 years.
- If the cholesterol is ≥240 mg/dL or if the cholesterol is between 200 and 230 mg/dL with an HDL-c <35 mg/dL and two or more other risk factors, a formal lipoprotein analysis should be done.

A lipoprotein analysis requires a 12-hour fast in order to allow all chylomicrons and intermediate-density lipoproteins to be cleared from the circulation. With this accomplished, the very low-density lipoprotein (VLDL) cholesterol can be estimated. This estimation evolves from the relation of triglyceride to cholesterol in the VLDL particle of 5 mg triglyceride (TG) to 1 mg cholesterol; this gives rise to the Friedewald formula:

$$\text{LDL-c} = \text{Total cholesterol} - \text{HDL-c} - (\text{TG}/5)$$

The estimate of VLDL-c is accurate up to triglyceride levels of 400 mg/dL. Beyond this, the estimate of LDL-c with this formula is unreliable, and the LDL-c level cannot be calculated. The formula is also invalid in patients with the rare type III abnormality (familial dysbetalipoproteinemia).

Based on the results of this calculation in asymptomatic patients, therapy can be tailored to the LDL-c level and other risk factors (Table 5.4). According to the NCEP:

- If the LDL-c is <130 mg/dL, no further action is suggested. Cholesterol and HDL-c levels should be measured in 5 years.
- If the LDL-c is between 130 and 159 mg/dL and there are fewer than two risk factors, dietary advice and advice on physical activity should be given and the cholesterol should be rechecked in 1 year.
- If the LDL-c is between 130 and 159 mg/dL and there are two or more risk factors or if the LDL-c is ≥160 mg/dL, a thorough evaluation including evaluations for secondary causes and familial disorders, is warranted.

5

In patients with diagnosed atherosclerotic disease (myocardial infarction [MI], angina pectoris, peripheral vascular or cerebrovascular diseases), the guidelines for screening are more strict. In these patients, a lipoprotein analysis should be undertaken. This analysis should not be done within 6 weeks from hospital discharge, since the effects of illness may lead to underestimation of the LDL-c level. The NCEP provides specific advice about when to *initiate LDL–c-lowering therapy*. In general, a 6-month trial of intensive dietary therapy should precede use of an LDL-c lowering medication. Much briefer trials with diet as sole therapy are appropriate for patients with established atherosclerotic disease or for those with marked elevations of LDL-c (>220 mg/dL). Some experts in the field believe that this latter group of patients should be started on LDL–c-lowering drugs at the same time as advice on diet and exercise is given. Drug therapy could be reduced if LDL-c falls too far below 100 mg/dL with this regimen. However, the NCEP does not actually recommend such an aggressive approach.

TABLE 5.4 — NATIONAL CHOLESTEROL EDUCATION PROGRAM GUIDELINES FOR INITIATION OF LDL–C-LOWERING THERAPY

Patient Category	LDL-c Level	Therapy to Initiate	Goal Level (mg/dL)
Without CHD			
<2 risk factors (low risk)	≥160	Dietary	<160
	≥190	Drug	<160
≥2 risk factors (high risk)	≥130	Dietary	<130
	≥160	Drug	<130
With CHD (highest risk)			
	≥100	Dietary	<100
	≥130	Drug	<100

Abbreviations: LDL-c, low-density lipoprotein cholesterol; CHD, coronary heart disease.

Expert Panel on Detection, Evaluation, and Treatment of High Blood Cholesterol in Adults. *JAMA*. 1993;269:3020.

The use of LDL–c-lowering drugs is then considered for those in the lowest risk group (fewer than two risk factors and no established atherosclerotic disease) if LDL-c remains >190 mg/dL despite nonpharmacologic efforts. In the intermediate-risk group (two or more risk factors and no established atherosclerotic disease), drug therapy is considered if LDL-c remains >160 mg/dL.

The NCEP guidelines do allow for the use of clinical judgment in using LDL–c-lowering drugs for patients with CHD and LDL-c levels in the range of 101 to 130 mg/dL. Interestingly, in the Cholesterol and Recurrent Events (CARE) study where patients were enrolled 3 to 20 months post-MI, the group with initial LDL-c <125 mg/dL did not show benefit with treatment as contrasted with the placebo group (Table 5.5). On the other hand, in the Lipoprotein and Coronary Atherosclerosis Study (LCAS), equal angio-

TABLE 5.5 — EFFECT OF PRAVASTATIN ON MYOCARDIAL INFARCTION SURVIVORS

Coronary Event	Pravastatin (%)	Placebo (%)
MI or death	10.2%	13.2%
CABG	7.5%	10.0%
PTCA	8.3%	10.5%

Pravastatin administered to 4159 patients who survived MI and had cholesterol <240 mg/dL (with average LDL 139 mg/dL); reduction in fatal or nonfatal MI was shown along with a reduction in the need for subsequent invasive therapy.

Abbreviations: MI, myocardial infarction; CABG, coronary artery bypass graft; PTCA, percutaneous transluminal coronary angioplasty; LDL, low-density lipoprotein.

Sacks FM, et al. *N Engl J Med*. 1996;335:1001-1009.

graphic benefit was seen in those with LDL-c <130 mg/dL as compared with those with higher LDL-c (see Chapter 7, *Clinical Trials*).

In general, ratios of HDL-c and either LDL-c or total cholesterol are not addressed in any of the recommendations for evaluation of patients with hyperlipidemia because both the LDL-c and HDL-c should be considered individually. Thus, while ratios may in some instances be helpful in impressing patients with the need to modify lifestyle and perhaps to take medications, they are less useful in formulating a clinical approach to individual patients.

Evaluation of triglyceride levels is a major source of confusion for the practicing physician. Several caveats should be stressed:

- Triglycerides are a major component of chylomicrons and VLDL and intermediate-density lipoprotein particles.
- Triglyceride levels should be obtained on fasting plasma (10 to 12 hours should be sufficient in most cases).
- Severe elevations of triglyceride levels can be determined by observing a creamy supernatant in blood that has been chilled to 4°C in an upright test tube. The appearance of a dramatic creamy layer measuring one finger breadth or more in height suggests triglyceride values exceeding 1000 mg/dL. (A light topping of "cream" often means the patient is not fasting and that normal postprandial lipemia is being observed.)
- Triglyceride levels are highly variable. Unless scrupulous attention is given to weight, exercise, alcohol intake, and dietary composition, levels can vary widely, even in samples drawn within a few days of each other. Thus multiple measurements for triglycerides should be obtained to determine baseline levels.

- There is still variation as to what constitutes a normal triglyceride value. For those with diabetes, CHD, or atherogenic dyslipidemia (see Chapter 4, *Causes of Secondary Hyperlipoproteinemia*), it may be useful to lower the triglycerides below 150 mg/dL.

> ***Clinical Tip:*** There is one situation where a "non-fasting" triglyceride is helpful. In the patient with abdominal pain in the emergency room, creamy blood should suggest hyperlipidemic acute pancreatitis. A triglyceride value greatly in excess of 1000 mg/dL makes this diagnosis very likely.

Laboratory Testing

All laboratory tests have an intrinsic variability, and cholesterol tests are no exception. The two major areas of variability that must be addressed in the context of evaluation of patients for cholesterol-based risk and for effects of therapy are:

- Accuracy
- Precision.

Accuracy relates to how closely an individual test technique gives a result that is close to the lipid research clinics' method (the Abell-Kendall method for total cholesterol). A deviation from this value is termed "bias" (Table 5.6).

Precision (coefficient of variation—CV %) relates to how close individual values are to each other when tested within a single technique. In evaluating a testing technique, both factors must be addressed. A test may be very precise but very inaccurate (Table 5.7). In general, tests should be within 3% for both accuracy and precision. Even with this very narrow tolerance, the estimation of lipid levels in a single patient

TABLE 5.6 — ACCURACY OF CHOLESTEROL MEASUREMENT		
LRC Value	**SMAC**	**ACA**
170	190	200
200	225	240
220	250	265
240	275	290
260	295	315

These two commonly used analyzers demonstrate a consistent "bias" to higher values on a single sample.

Abbreviations: LRC, lipid research clinic; SMAC, Technicon SMAC (cholesterol oxidase ≥ quinone dye); ACA, Dupont ACA (cholesterol oxidase ≥ n,*N*-diethylaniline HCl).

may have a fairly wide variation. Therefore, before making a decision to initiate therapy, the clinician should be sure that the test will give a correct result. Before a recommendation of a major change in lifestyle or drug therapy is given, there should be at least two values within 10% of each other.

Clinical factors that must be considered when blood is drawn include the following:

- Is the patient fasting for 12 hours? If triglycerides are measured, the patient should be fasting; this allows calculation of LDL-c. On the other hand, for screening purposes, a nonfasting total cholesterol and HDL-c are allowed. Water and black coffee are allowed.
- Is the patient supine or upright? Supine patients sampled in the hospital have values for cholesterol that are 10% below those seen when they are tested as outpatients. This is due to a dilutional effect.

TABLE 5.7 — HOW LAB TESTS MAY NOT GIVE A TRUE ESTIMATE OF A LIPID LEVEL				
	Precision (CV %)		**Bias (≈ ReferenceValue)**	
True Value	**+ 3%**	**– 3%**	**+ 3%**	**– 3%**
200	206	194	212	188
210	216	204	223	198
220	227	213	233	207
230	237	223	255	226
240	247	233	254	226
250	258	242	265	235
260	268	252	276	245
270	278	262	286	254
280	288	272	297	263
290	299	283	308	273
300	309	291	318	282

Abbreviation: CV, coefficient of variation.

- Is the patient in a steady state? A recent dietary change or change in weight will make lipid values difficult to evaluate. Recent weight loss will improve and recent weight gain will exacerbate lipid values. Thus lipid values are best determined when weight and diet are steady for at least 3 weeks. In addition, cholesterol values can decline after major surgery, burns, or significant infections.

Clinical Tip: Patients should be told to discontinue lipid-lowering drugs when taking antibiotics for significant infections; this minimizes the chance for drug-drug interactions. Several instances of liver toxicity have emerged when patients have added antibiotics to their lipid-lowering regimens.

- Has the patient sustained an MI? Initial values obtained in the first 24 hours after MI can be used, but after that, triglycerides rise and cholesterol values fall. This situation has less practical significance now because patients with initial LDL-c down to 125 mg/dL appear to benefit from lipid-lowering therapy.

Outside of the United States, lipid values are reported in units (mmol/L) of the Système International (SI). In the United States, lipid values are reported in units (mg/dL) of the metric system. Table 5.8 gives representative values in both systems of measurement.

Goals of Therapy

One of the more important contributions of the NCEP has been defining goals of therapy, which allows physicians to tailor therapy for patients in an appropriate manner. Although those with known CHD who are candidates for secondary prevention have the highest priority for lowering of LDL-c, such intense therapy is not proper for younger patients who are not greatly at risk for CHD but whose cholesterol is elevated (Table 5.4):

- For those at low risk, the LDL-c goal is <160 mg/dL.
- For those at high risk, with two or more risk factors, the goal is <130 mg/dL.
- For those with known CHD, at highest risk, the goal is <100 mg/dL.

TABLE 5.8 — CONVERSION OF LIPID LEVELS BETWEEN MEASUREMENT SYSTEMS			
Cholesterol*		**Triglyceride†**	
mg/dL	**SI (mmol/L)**	**mg/dL**	**SI (mmol/L)**
240	6.2	1000	11.3
200	5.2	400	4.5
190	4.9	200	2.3
160	4.1	—	—
130	3.4	—	—
100	2.6	—	—
60	1.6	—	—
35	0.9	—	—

Abbreviations: SI; International System of Units.

* To convert cholesterol, the factor is 38.8.
† To convert triglycerides, the factor is 88.5.

Table 5.4 shows how treatment decisions are guided by both LDL-c level and clinical status. As emphasized throughout this book, clinical stratification is crucial because those patients with established atherosclerotic vascular disease deserve the highest priority for intensive lowering of their LDL-c.

Recently, the National Committee for Quality Assurance (NCQA) implemented a performance measure as part of the Health Plan Employer and Data Information Set (HEDIS) for LDL-c (Lee et al, 2000). The new HEDIS measure will require managed care organizations seeking NCQA accreditation to measure and report the percentage of patients with major CHD events who achieve LDL-c levels <130 mg/dL between 60 and 365 days after discharge. The threshold for the performance measure was set higher than the clinical goal for several reasons. This was because the large randomized statin trials have not specifically addressed the question of whether lowering LDL-c from 100 to 129 mg/dL to below 100 mg/dL will yield substantially improved clinical outcomes (although such trials are currently under way). Also, it was expected that greater than average reductions in LDL-c would be required to reach 100 mg/dL or less in high-risk populations. In fact, in the Scandinavian Simvastatin Survival Study (4S), the CARE trial, and the Long-Term Intervention With Pravastatin in Ischemic Disease (LIPID) trial, less than half of patients overall achieved an LDL-c goal of less than 100 mg/dL while taking average doses of statins. Although intensification of therapy is feasible, the NCQA did not want the performance measure to mandate that physicians intensify drug therapy for those patients whose LDL-c levels were between 100 and 129 mg/dL and instead wished to emphasize individualized clinical judgment.

SUGGESTED READINGS

Blank DW, Hoeg JM, Kroll MH, Ruddel ME. The method of determination must be considered in interpreting blood cholesterol levels. *JAMA*. 1986;256:2867-2870.

Expert Panel on Detection, Evaluation, and Treatment of High Blood Cholesterol in Adults. Summary of the second report of the National Cholesterol Education Program (NCEP) Expert Panel on Detection, Evaluation, and Treatment of High Blood Cholesterol in Adults (Adult Treatment Panel II). *JAMA*. 1993;269:3015-3023.

Herd JA, Ballantyne CM, Farmer JA, et al. Effects of fluvastatin on coronary atherosclerosis in patients with mild to moderate cholesterol elevations (Lipoprotein and Coronary Atherosclerosis Study [LCAS]). *Am J Cardiol*. 1997;80:278-286.

Lee TH, Cleeman JI, Grundy SM, et al. Clinical goals and performance measures for cholesterol management in secondary prevention of coronary heart disease. *JAMA*. 2000;283:94-98.

Manson JE, Tosteson H, Ridker PM, et al. The primary prevention of myocardial infarction. *N Engl J Med*. 1992;326:1406-1416.

Muldoon MF, Bachen EA, Manuck SB, Waldstein SR, Bricker PL, Bennett JA. Acute cholesterol responses to mental stress and change in posture. *Arch Intern Med*. 1992;152:775-780.

Pais P, Pogue J, Gerstein H, et al. Risk factors for acute myocardial infarction in Indians: a case-control study. *Lancet*. 1996; 348:358-363.

Sacks FM, Pfeffer MA, Moye LA, et al. The effect of pravastatin on coronary events after myocardial infarction in patients with average cholesterol levels. Cholesterol and Recurrent Events Trial investigators. *N Engl J Med*. 1996;335:1001-1009.

6 Atherosclerosis, Lipids, and Coronary Syndromes

Depending on the genetic predisposition and constellation of risk factors (see Chapter 5, *Clinical Appraisal and Goals of Therapy*), clinically important atherosclerotic lesions are generally seen after the third to fourth decade of life. However, as recently as 1993, moderate atherosclerosis was seen in the coronary arteries of more than 75% of asymptomatic young men who died of trauma. Atherosclerosis is due to cholesterol deposition in the wall of arteries. The process is initiated by low-density lipoprotein cholesterol (LDL-c) migrating into the subendothelial region of an arterial wall, either because the concentration of LDL-c in the blood is so high that it can get into the subendothelial region by a diffusion-like mechanism and/or there is injury to the endothelial cells. The cells can be injured due to:

- Physical trauma (blood pressure)
- Toxins (tobacco products)
- Some infections (some viruses or *Chlamydia* pneumonia)
- Immune complexes.

The LDL in the subendothelial interstitial space becomes oxidized (modified). During the process of becoming oxidized, the LDL is initially only partly oxidized and develops properties crucial to the atherosclerotic process. The modified LDL is then taken up by cells (monocytes/macrophages and/or smooth-muscle cells) that become engorged to form cholesterol-laden foam cells (type I, II, and II lesions). This may set into play a complex process involving production of various proteins not native to the normal vascular wall, including:

- Growth factors
- Cytokines
- Proteins involved in calcium deposition
- Vasoregulatory molecules (see *Suggested Readings*) (Table 6.1).

Primary cytokines also stimulate production of interleukin-6, which induces expression of hepatic genes leading to a rise in acute-phase reactants found in blood, including C-reactive protein (CRP) and serum amyloid-A. Using ultrasensitive techniques, levels of CRP, for example, can provide a glimpse into the intensity of the underlying inflammation in the artery wall (Libby and Ridker, 1999).

The beginning of atherosclerosis is seen in childhood, with minor deposition of cholesterol in the subintimal layers of arteries, generally at points of stress, including:

- Bifurcations
- Branch points of vessels.

The American Heart Association working group on classification of coronary lesions (Table 6.2) labels these type I-III lesions, none of which are occlusive. In fact, it is probable that type I, and even type II, lesions may appear and then undergo spontaneous regression (ie, they may "come and go"). Many of these lesions will progress to more complex lesions. It is important to note that the rate of progression of atherosclerotic lesions is neither uniform nor predictable (Figure 6.1). The so-called complex lesions are types IV, V, and VI. For practical purposes, a type VI lesion is any lower-grade lesion that develops fissuring and superimposed thrombus.

There are three major clinical coronary syndromes:

- Sudden death

TABLE 6.1 — FUNCTIONS AND EFFECTS OF CELLS INVOLVED IN ATHEROSCLEROSIS	
Type of Cell	**Major Functions and Effects**
Endothelial cells	Inhibits platelet aggregation and clotting; governs normal vascular tone (nitric oxide [NO] and endothelin)
Platelets	Elaborate growth factors; induce clotting; release factors to promote vascular "spasm"
Macrophages	Take up modified low-density lipoprotein cholesterol (LDL-c); become "foam cells"; may elaborate proteins that participate in modeling; monocyte chemotactic protein-1; tumor necrosis factor; osteopontin, osteonectin, osteocalcin; may release proteases with cell injury or death
Smooth-muscle cells	Take up LDL-c and become "foam cells"; secrete mitogenic cells factors; generate basement membrane proteins
Lymphocytes	Modify actions of macrophages; elaborates matrix metalloproteinases
Extracellular matrix: Cholesterol and other lipids Proteoglycans Collagen Elastin Calcium	Generally the result of the above functions; is a dynamic biologically active mileau

- Myocardial infarction (MI)
- Angina pectoris (stable or unstable).

All three of the clinical syndromes are related to coronary atherosclerosis. However, the syndromes of un-

TABLE 6.2 — CLASSIFICATION OF ATHEROSCLEROTIC ARTERIAL LESIONS

Type of Lesion	Common Term	Components
I	Adaptive intimal thickening	Atherogenic lipoprotein, scattered macrophage foam cells, lymphocytes
II	Fatty streak	Macrophage-derived foam cells; lymphocytes; lipid-laden smooth–muscle-derived foam cells
III	—	Macrophage foam cells; lymphocytes; lipid-laden smooth-muscle foam cells; calcium
IV	Atheroma	Extracellular lipid
V	Fibrous plaque	New fibrous connective tissue
Va	Fibroatheroma	—
Vb	Calcific lesion	Diffuse, more severe calcium; fibrous connective tissue
Vc	Fibrotic lesion	Minimal lipid
VI	Complex lesion	Thrombus; hematoma; surface defect

Stary HC, et al. *Arterioscler Thromb Vasc Biol.* 1995;15:1512-1531.

FIGURE 6.1 — SCHEMA OF PROGRESSION OF STAGES IN ATHEROSCLEROSIS

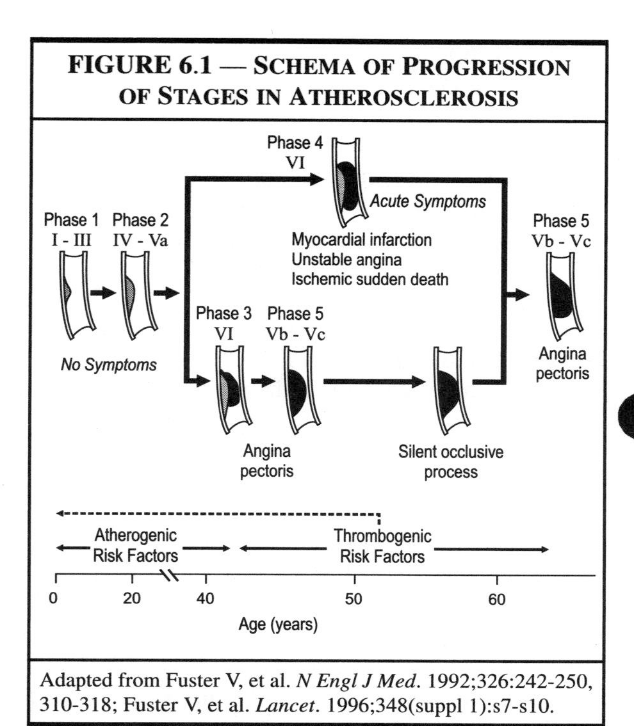

Adapted from Fuster V, et al. *N Engl J Med*. 1992;326:242-250, 310-318; Fuster V, et al. *Lancet*. 1996;348(suppl 1):s7-s10.

stable angina pectoris and MI have a different vascular pathophysiology from that of chronic stable angina. All three are caused by an imbalance between myocardial oxygen supply and demand.

Chronic stable angina is precipitated by an increase in myocardial oxygen demand (MVO_2) when there is critical (usually thought to be greater than 70%) fixed stenosis of one or more of the coronary arteries. These stenoses tend to be stable but severe (type Vb, Vc). The increased MVO_2 is brought on by factors that will increase cardiac work, such as:

- Exercise
- Mental stress
- Anemia

- Infection
- Emotional stress.

Unstable angina and MI are usually brought on by:
- Rapid progression of a previously less important (<50%) narrowing
- Sudden (partial or total) occlusion of a coronary artery by a thrombus.

This latter progression most often occurs on complex lesions. Complex lesions generally:
- Are more eccentric in the vessel
- Have a thinner cap
- Have a more cholesterol-rich core
- May be more immature.

The vulnerable plaque is usually moderate (not severe) in size and has a thin fibrous cap with a large lipid pool. These caps frequently fissure and then break open, exposing the blood in the vessel lumen to the cholesterol and collagen content of the plaque. This condition will then lead to initiation of the clotting cascade over the site of the plaque fissure, potentially resulting in thrombotic occlusion of the involved vessel. When the vessel suddenly becomes occluded, myocardial cell damage occurs. If the occlusion is subtotal or intermittent (ie, if spontaneous thrombolysis occurs), the syndrome is usually unstable angina pectoris (resting pain with transient electrocardiogram [ECG] changes and possibly minimal creatinine phosphokinase-MB, or troponin release).

If the occlusion is sudden and total, MI will follow. The extent of the infarction depends on the amount of myocardium that has been deprived of blood (more severe with proximal vessel occlusions or in multivessel coronary artery disease, where other vessels cannot provide collateral flow) and the duration of occlusion. The prompt application of reperfusion, either with thrombolytic therapy or percutaneous transluminal coronary angioplasty (PTCA), can limit the extent of infarction.

In addition, progression frequently occurs in lesions that are less severe than in those that are more severe. These slowly developing, complete occlusions may cause myocardial infarct or unstable angina but may frequently be silent. Approximately 70% of coronary occlusions responsible for the syndrome of unstable angina had less than a 50% stenosis on an angiogram performed some time before the episode. The culprit lesion of MI was also frequently less than 50% occluded in over two thirds of the patients and less than 70% occluded in over half. Thus it is not surprising that functional tests such as the exercise treadmill test, with or without nuclear perfusion imaging (which depends on critical coronary stenosis), cannot predict all patients who will subsequently have an MI (Figure 6.2).

Studies on the effect of vigorous reduction of cholesterol on the atherosclerotic process have generally shown a reduction of the incidence of MI, death and the need for coronary artery bypass graft (CABG) or PTCA (Table 6.3). Other studies have shown a mild reduction in plaque size and demonstrated slower progression of lesions. In all the regression studies, however, the reduction in event rates (MI, unstable angina, need for CABG or PTCA) is more than would be expected by the amount of change in lumen diameter (Table 6.4). This suggests that stabilization of mild to moderate, as yet asymptomatic, plaques accomplished with reduction of cholesterol is as important or more so than any benefit due to regression of plaque size.

FIGURE 6.2 — STENOSIS SEVERITY AND ASSOCIATED RISK OF MYOCARDIAL INFARCTION

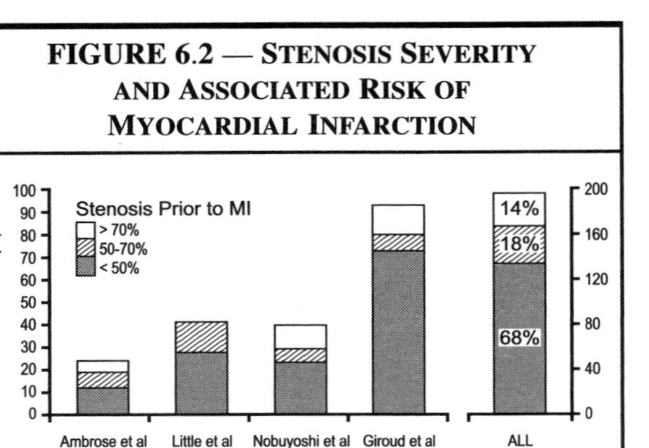

Abbreviations: MI, myocardial infarction.

Falk E, et al. *Circulation.* 1995;92:657-671.

SUGGESTED READINGS

Creager MA, Selwyn A. When "normal" cholesterol levels injure the endothelium. *Circulation.* 1997;96:3255-3257. Editorial.

Falk E, Shan PK, Fuster V. Coronary plaque disruption. *Circulation.* 1995;92:657-671.

Fuster V, Badimon L, Badimon JJ, Chesebro JH. The pathogenesis of coronary artery disease and the acute coronary syndromes. *N Engl J Med.* 1992;326:242-250, 310-318.

Fuster V, Fallon JT, Nemerson Y. Coronary thrombosis. *Lancet.* 1996;348(suppl 1):s7-s10.

Joseph A, Ackerman D, Talley JD, Johnstone J, Kupersmith J. Manifestations of coronary atherosclerosis in young trauma victims—an autopsy study. *J Am Coll Cardiol.* 1993;22:459-467.

Jost S, Deckers JW, Nikutta P, et al. Evolution of coronary stenoses is related to baseline severity—a prospective quantitative angiographic analysis in patients with moderate coronary disease. INTACT Investigators. International Nifedipine Trial on Antiatherosclerotic Therapy [published erratum appears in *Eur Heart J.* 1995;16:293]. *Eur Heart J.* 1994;15:648-653.

TABLE 6.3 — SELECTED PREVENTION STUDIES

Study					
Name	Type	*n*	Average LDL Reduction (%)	Mean Decrease in Fatal or Nonfatal MI (Range) (%)	Procedure Decrease (%)
ACAPS	Primary prevention	918	28	> 50 (from 2 to 0.2)	—
AFCAPS/TexCAPS	Primary prevention	6605	25	26 (from 54 to 8)	33
CARE	Secondary prevention	4159	28	24 (from 13 to 10)	27
4S	Secondary prevention	4444	35	25 (from 11 to 8.2)	34
LIPID	Secondary prevention	9014	25	23	—
PLAC I & II	Secondary prevention	559	27 to 28	>50	—
REGRESS	Secondary prevention (regression)	885	29	42	57
WOSCOPS	Primary	6595	26	31 (from 7.9 to 5.5)	37

A more complete review of angiographic and clinical lipid-lowering trials is in Chapter 7, *Clinical Trials*.

Abbreviations: *n*, number of patients; LDL, low-density lipoprotein; MI, myocardial infarction; ACAPS, Asymptomatic Carotid Artery Progression Study; AFCAPS/TexCAPS, Air Force/Texas Coronary Atherosclerosis Prevention Study; CARE, Cholesterol and Recurrent Events Trial; 4S, Scandinavian Simvastatin Survival Study; LIPID, Long-Term Intervention with Pravastatin in Ischemic Disease; PLAC, Pravastatin, Lipids and Atherosclerosis in the Carotid Artery; REGRESS, Regression Growth Evaluation Statin Study; WOSCOPS, West of Scotland Coronary Prevention Study.

TABLE 6.4 — REDUCTION OF EVENT RATES IN SELECTED REGRESSION STUDIES*		
Study	**Change in Average Stenosis (%)**	**Change in CV Event Rate (%)**
FATS	– 0.8	– 73
LCAS	– 2.2	– 24†
MAAS	– 2.67	– 22†
SCRIP	– 0.8	– 50
STARS	– 7	– 79

Abbreviations: CV, cardiovascular; FATS, Familial Atherosclerosis Treatment Study; LCAS, Lipoprotein and Coronary Atherosclerosis Study; MAAS, Multicentre Anti-Atheroma Study; SCRIP, Stanford Risk Intervention Project; STARS, St. Thomas Atherosclerosis Regression Study.

* See Chapter 7, *Clinical Trials,* for further details of the individual studies.

† Not statistically significant.

Lee RT, Libby P. The unstable atheroma. *Arterioscler Thromb Vasc Biol.* 1997;17:1859-1867.

Lewis B. Coronary heart disease prevention and the infarctogenic plaque. *Isr J Med Sci.* 1996;32:360-363.

Libby P, Ridker PM. Novel inflammatory markers of coronary risk theory versus practice. *Circulation.* 1999;100:1148-1150. Editorial.

Little WC. Angiographic assessment of the culprit coronary artery lesion before acute myocardial infarction. *Am J Cardiol.* 1990;66: 44G-47G.

Manson JE, Tosteson H, Ridker PM, et al. The primary prevention of myocardial infarction. *N Engl J Med.* 1992;326:1406-1416.

Ross R. The pathogenesis of atherosclerosis: a perspective for the 1990s. *Nature.* 1993;362:801-809.

Ross R. The pathogenesis of atherosclerosis–an update. *N Engl J Med.* 1986;314:488-500.

Stary HC, Chandler AB, Dinsmore RE, et al. A definition of advanced types of atherosclerotic lesions and a histological classification of atherosclerosis. A report from the Committee on Vascular Lesions of the Council on Arteriosclerosis, American Heart Association. *Arterioscler Thromb Vasc Biol.* 1995;15:1512-1531.

Steering Committee of the Physicians' Health Study Research Group. Final report on the aspirin component of the ongoing Physicians' Health Study. *N Engl J Med.* 1989;321:129-135.

Wexler L, Brundage B, Crouse J, et al. Coronary artery calcification: pathophysiology, epidemiology, imaging methods, and clinical implications. A statement for health professionals from the American Heart Association. Writing Group. *Circulation.* 1996;94:1175-1192.

7 Clinical Trials

Cholesterol lowering, along with smoking cessation, beta-blockers, and aspirin and other antiplatelet agents, has become a mainstay of secondary prevention of coronary heart disease (CHD). For selected patients, angiotensin-converting enzyme inhibitors, estrogens, blood pressure control, and weight management are other measures to be added as well. Newer information also suggests that in selected groups at high risk, primary prevention with more aggressive lipid lowering may also be beneficial. This chapter will provide a summary of the compelling data supporting these new patterns of clinical care.

Prospective clinical trials using changes seen on serial coronary angiography have provided many of the insights demonstrating the crucial impact of aggressive lipid lowering. In the past 25 years, these trials have looked at coronary, femoral, and with the use of ultrasound, carotid atherosclerosis. Lipid-lowering methods include the use of single lipid-lowering drugs; multiple drugs; multiple forms of intervention, including diet, exercise, stress management, and even surgical approaches. Figure 7.1 shows the wide range of low-density lipoprotein cholesterol (LDL-c) levels encompassed by angiographic trials of cholesterol lowering.

Angiographic Trials Using Monotherapy

■ Resins

The first group of trials to discuss looked at the use of bile acid sequestrants, such as cholestyramine

FIGURE 7.1 — REGRESSION STUDIES EMPLOYING ANGIOGRAPHY

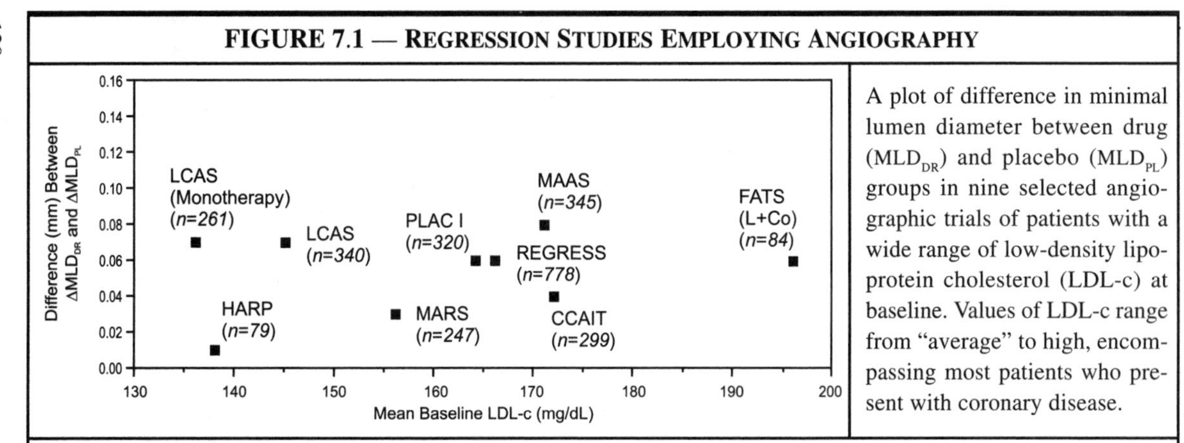

A plot of difference in minimal lumen diameter between drug (MLD_{DR}) and placebo (MLD_{PL}) groups in nine selected angiographic trials of patients with a wide range of low-density lipoprotein cholesterol (LDL-c) at baseline. Values of LDL-c range from "average" to high, encompassing most patients who present with coronary disease.

Abbreviations: LCAS, Lipoprotein and Coronary Atherosclerosis Study; PLAC 1; Pravastatin Limitation of Atherosclerosis in the Coronary Arteries; MAAS, Multicentre Anti-Atheroma Study; FATS, Familial Atherosclerosis Treatment Study; REGRESS, Regression Growth Evaluation Statin Study; HARP, Harvard Atherosclerosis Reversibility Project; MARS, Monitored Atherosclerosis Regression Study; CCAIT, Canadian Coronary Atherosclerosis Intervention Trial.

Adapted from Herd JA, et al. *Am J Cardiol*. 1997;80:278-286.

or colestipol (Table 7.1). Cholestyramine was used in the Type II Coronary Intervention Trial of the National Heart, Lung, and Blood Institute (NHLBI). This trial suffered from small numbers and a long recruitment phase. Eventually 116 men and women were recruited. Diet and doses of cholestyramine up to 24 g/d resulted in a 26% fall in LDL-c and an 8% rise in high-density lipoprotein cholesterol (HDL-c) values. Panels of blinded angiographers found that coronary artery disease progressed in 49% (28 of 57) of the placebo-treated patients vs 32% (19 of 59) of the cholestyramine-treated patients ($P < 0.05$). Of lesions causing 50% or more narrowing in a coronary vessel, 33% of placebo-treated and 12% of cholestyramine-treated patients manifested lesion progression ($P < 0.05$).

The St Thomas' Atherosclerosis Regression Study (STARS) had three arms. The control group was a usual-care arm and it was compared with both a diet-only arm and an arm with diet and cholestyramine at a dosage of 16 g/d. These 90 men were hypercholesterolemic, with mean total cholesterol values of 280 mg/dL. Using computerized angiographic analysis, investigators saw improvement both in the primary end point of mean absolute width of the arterial segments for both the diet-only and the diet-and-cholestyramine groups as compared with usual-care groups. Of interest, there were significantly more clinical events in the usual-care group than in either the diet-plus-cholestyramine or the diet-only arm.

Subsequent analysis showed that angiographic progression was directly, strongly, and independently associated with intake of saturated fatty acids; this included stearic acid, which is not cholesterol raising. This effect was not fully explained by the influence of saturated fat in raising serum cholesterol; after adjustment for LDL-c level, stearic acid intake remained independently predictive of progression. Furthermore,

TABLE 7.1 — RESIN TRIALS

Study	*n*	Years	Gender	Cholestyramine Dose	% Δ LDL	% Δ HDL	Comment
NHLBI	116	5	Men and women	16 g/d (average)	– 26	8	Significant effect in lesions >50% at baseline
STARS	26*	2	Men	16 g/d	– 35.7	4	Improvement in angiography and clinical status

Abbreviations: *n*, number of patients; LDL, low-density lipoprotein; HDL, high-density lipoprotein; NHLBI, National Heart, Lung, and Blood Institute's Type II Coronary Intervention Trial; STARS, St Thomas' Atherosclerosis Regression Study.

* The 26 men represented one of the three groups from the STARS trial—those assigned to cholestyramine resin.

the intake of trans fatty acids was directly related to progression. There was no protective effect seen from either n-6 or n-3 fatty acids.

HMG-CoA Reductase Inhibitors

■ Lovastatin

The two trials which used lovastatin alone were the Monitored Atherosclerosis Regression Study (MARS) and the Canadian Coronary Atherosclerosis Intervention Trial (CCAIT). Both of these studies looked at men and women with angiographic coronary disease and elevated cholesterol levels in approximately the 200 to 300 mg/dL range. There was a wide range of coronary status in MARS at baseline with both one-vessel and severe three-vessel disease. Quantitative coronary angiography did not show significant changes between the lovastatin and placebo groups. On the other hand, the blinded panel readings showed important differences between treated and control patients. For example, fewer than half of the group treated with lovastatin showed progression as compared with approximately two thirds of the group given placebo. Regression was seen in 23% of those receiving treatment vs only 11% in the placebo group. Severe progression likely to require revascularization was seen more often in the placebo group.

A follow-up paper looked at the relationships between lipoprotein subclasses and coronary progression in MARS. The analyses showed that the mass of the smallest LDL, all very low-density lipoprotein (VLDL), and peak LDL flotation rates were significantly related to the progression of coronary lesions, specifically the low-grade lesions. Greater baseline levels of HDL_3, a subclass of HDL, were related to a lower likelihood of coronary lesion progression. Finally, further analysis by the MARS investigators showed that lesion progression in one coronary segment was associated with significant increases in the arterial diameter of remote segments. In addition, there was a trend, although not significant, for a decrease in arterial diameter in remote coronary segments in response to lesion regression. The authors felt that this was most likely the result of flow-mediated vascular compensation, which was enhanced by lovastatin's lowering of apo B and C-III levels.

In the CCAIT, there was less progression and fewer new lesions in the lovastatin-treated group as compared with the placebo group. There was a trend toward fewer coronary events, but this was not statistically significant. Interestingly, the coronary change score for those taking lovastatin in the subjects with LDL-c above the median of 176 mg/dL was significantly better than that seen with the placebo (Table 7.2).

In the Asymptomatic Carotid Artery Progression Study (ACAPS), lovastatin was shown to significantly affect the progression of intimal-medial thickness (IMT) in a 3-year study. In addition, in this asymptomatic group of men and women, there were strikingly fewer cardiovascular endpoints (5 vs 14) and deaths (1 vs 8) in the lovastatin-treated group than in the placebo group.

■ Pravastatin

Clinical data from four regression trials were pooled for a predetermined analysis of the effect of pravastatin therapy on the risk of coronary events. The mean LDL-c for the four populations ranged from 163 to 189 mg/dL. There was a 62% reduction in fatal and nonfatal myocardial infarction (MI) in those in the pravastatin group. This effect was seen in older and younger subjects, men and women, and subjects with and without histories of either hypertension or previous MI. There was a nonsignificant 46% reduction in all-cause mortality. There was also a 62% re-

TABLE 7.2 — LOVASTATIN TRIALS

Study	n	Years	Gender	Lovastatin Dose (mg)	% Δ LDL	% Δ HDL	Comment
ACAPS	919	3	Men and women	26	– 28	5	Lovastatin group significantly reversed the progression of carotid intimal-medial thickness; also, fewer morbid and fatal cardiovascular events
CCAIT	331	2	Men and women	36	– 29	7.3	Benefit from treatment seen in lesions >50%
MARS	270	2	Men and women	80	– 38	8.5	Lesions >50% at baseline improved with therapy; HDL_3 was protective; also, lovastatin significantly reversed progression of common carotid intimal-medial thickness

Abbreviations: *n*, number of patients; LDL, low-density lipoprotein; HDL, high-density lipoprotein; ACAPS, Asymptomatic Carotid Artery Progression Study; CCAIT, Canadian Coronary Atherosclerosis Intervention Trial; MARS, Monitored Atherosclerosis Regression Study.

7

duction in the risk of fatal or nonfatal stroke. Of interest is the apparent benefit among patients with lower LDL-c levels. The multinational study looked at hypercholesterolemic subjects with coronary disease and/or multiple risk factors. The marked reduction in clinical events was unexpected (Table 7.3).

Finally, the Kuopio Atherosclerosis Prevention Study (KAPS); Pravastatin, Lipids, and Atherosclerosis in the Carotid Artery (PLAC-II); and the Carotid Atherosclerosis Italian Ultrasound Study (CAIUS)—using pravastatin—joined the Asymptomatic Carotid Artery Progression Study (ACAPS) and MARS, using lovastatin, in showing that statins which lower LDL-c significantly can effectively control the progression of carotid IMT as well as have a positive impact on morbidity and mortality from CHD. This further validates the use of statins as a class to reduce both CHD and stroke in high-risk patients.

■ Simvastatin

The Multicentre Anti-Atheroma Study (MAAS) was a multicenter trial in Europe which looked at 20 mg of simvastatin daily vs placebo. In this trial, 381 patients, of whom 88% were males, were followed for 4 years and then repeat angiography was performed. LDL-c was lowered 31% and there was beneficial angiographic change in the simvastatin group on measures of diffuse and focal coronary atherosclerosis. Importantly, there were significantly more new lesions and new total occlusions in the placebo group. There were no fewer deaths or MIs in the simvastatin group than in the placebo group, but the placebo group underwent revascularization procedures more often. The treatment effect was greater in segments with diameter stenosis of 50% or more at baseline. In those patients who had follow-up angiograms at 2 and 4 years, there was less incremental improvement in minimum lumen diameter and diameter stenosis at 4 years as compared with the gains at 2 years (Table 7.4).

■ Fluvastatin

The Lipoprotein and Coronary Atherosclerosis Study (LCAS) was important because it demonstrated the value of LDL lowering in coronary patients whose average LDL-c levels were lower than those looked at in previous angiographic trials. Approximately 25% of LCAS patients had baseline LDL-c levels below 130 mg/dL.

There was significantly less lesion progression and fewer new lesions in fluvastatin-treated patients than in controls. Importantly, benefit was seen whether the LDL-c was either above 160 mg/dL or below 131 mg/dL. Clinical significance was further amplified by showing treatment-related beneficial trends in myocardial perfusion as measured by positron-emission tomography. This study was large enough to permit subgroup analysis. An important subgroup of 84 patients had LDL-c <130 mg/dL. Even in these patients, the benefit from treatment seen on angiography was similar to that seen in those with higher LDL-c. Although there were consistent trends toward benefit with fluvastatin regarding clinical events or the need for further revascularization, these did not reach statistical significance. In this study, there were no cases of symptomatic liver disease or muscle disease due to fluvastatin. Among the fluvastatin-treated patients in the study group, levels of alanine transaminase or aspartate transaminase were elevated (above 3 times the upper limit of normal) in two and three patients, respectively. One patient in the placebo group had similarly elevated enzyme levels; no patients were given treatment. The only symptomatic patient had hepatitis B (Table 7.5).

TABLE 7.3 — PRAVASTATIN TRIALS

Study	***n***	**Years**	**Gender**	**Pravastatin Dose** (mg/d)	**% Δ LDL**	**% Δ HDL**	**Comment**
CAIUS	305	3	Men and women	40	– 23	3	Progression of carotid intimal-medial thickness significantly affected by pravastatin
KAPS	447	3	Men	40	– 27.4	– 1.9	Ultrasound-determined intimal-medial thickness improved by 45% with pravastatin
Multi-national	1062	1/2	Men and women	26.5 (average dose)	– 26	7	Six MIs, 5 cases unstable angina, 1 sudden death in placebo group; none in the pravastatin group

PLAC-I	408	3	Men and women	40	– 28	7	Reduced angiographic progression and clinical events of CHD
PLAC-II	151	3	Men and women	20 to 40 (73% of patients received 40 mg/d)	– 28	– 4	Efficacy of pravastatin therapy confined to common carotid
REGRESS	885	2	Men	40	– 28.2	10	Less progression of coronary atherosclerosis in treatment group

Abbreviations: *n*, number of patients; LDL, low-density lipoprotein; HDL, high-density lipoprotein; CAIUS, Carotid Atherosclerosis Italian Ultrasound Study; KAPS, Kuopio Atherosclerosis Prevention Study; MI, myocardial infarction; PLAC, Pravastatin, Lipids, and Atherosclerosis in the Carotid Artery; CHD, coronary heart disease; REGRESS, Regression Growth Evaluation Statin Study.

TABLE 7.4 — SIMVASTATIN TRIALS

Study	***n***	**Years**	**Gender**	**Simvastatin Dose** (mg/d)	**% Δ LDL**	**% Δ HDL**	**Comment**
MAAS	381	4	Men and women	20	– 31	9	Significantly fewer new lesions or total occlusions in treatment group

Abbreviations: *n*, number of patients; LDL, low-density lipoprotein; HDL, high-density lipoprotein; MAAS, Multicentre Anti-Atheroma Study.

TABLE 7.5 — FLUVASTATIN TRIALS

Study	*n*	Years	Gender	Fluvastatin Dose	% Δ LDL	% Δ HDL	Comment
LCAS	429	2	Men and women	20 mg bid; for those with higher LDL-c, fluvastatin + cholestyramine, 20 mg bid, as therapy	– 22.5	8.7	Signficantly less lesion progression and fewer new lesions in treated patients with improvement in myocardial blood flow by PET
LCAS*	84	2	Men and women	Fluvastatin, 20 mg bid, as monotherapy	– 27.9	8.0	Similar angiographic improvement with treatment

Abbreviations: *n*, number of patients; LDL, low-density lipoprotein; HDL, high-density lipoprotein; LCAS, Lipoprotein and Coronary Atherosclerosis Study; PET, positron emission tomography.

* Refers to a subgroup of LCAS with low-density lipoprotein cholesterol (LDL-c) <130 mg/dL; mean LDL-c of 121 mg/dL.

7

■ Partial Ileal Bypass

The Program on the Surgical Control of Hyperlipidemia (POSCH) was a trial of LDL-c lowering in subjects who had survived an MI and whose LDL-c level was >140 mg/dL. The partial small-bowel bypass operation is a less extensive and generally better tolerated operation than the more extensive and poorly tolerated ileal bypass operation that was used in the past for obesity. Nonetheless, an increase in stool frequency is commonly seen, and occasionally the partial ileal bypass must be reversed. This trial had the advantage of obtaining serial angiograms at 1, 5, 7, and 10 years after surgery. After 5 years, approximately 66% of the control group and only 33% of the treatment group showed progression of atherosclerosis. At 10 years, 85% of the control group showed progression as compared with 54.7% of the surgery group ($P < 0.001$). Overall, this obligate LDL-c lowering was associated with significantly fewer clinical events. The POSCH investigators were able to show that angiographic progression led to an increased rate of clinical events (Table 7.6).

Angiographic Trials Using Multiple Interventions

■ Double-Drug Trials

The use of combined therapy has several advantages. It takes advantage of the complementary action of the two drugs. It may be particularly useful either in patients with severe hyperlipidemia or in those with combined forms of hyperlipidemia where one drug may not be adequate (see Chapter 10, *The Lipid-Lowering Drugs*).

The Cholesterol Lowering Atherosclerosis Study (CLAS-I) was the first major trial to show convincingly that aggressive lipid-lowering therapy could cause definite angiographic improvement. This study

TABLE 7.6 — SURGERY TRIAL

Study	*n*	Years	Gender	Intervention	% Δ LDL	% Δ HDL	Comment
POSCH	838	5	Men and women	Partial ileal bypass surgery	– 42	5	Improvement in angiographic and clinical end points

Abbreviations: *n*, number of patients; LDL, low-density lipoprotein; HDL, high-density lipoprotein; POSCH, Program on the Surgical Control of Hyperlipidemia.

employed panels of blinded angiographers. The results were striking; in this group of nonsmoking, middle-aged men who had undergone coronary artery bypass graft (CABG) surgery, double-drug therapy with niacin and colestipol resin resulted in decreased progression and increased regression in native arteries and less new vessel formation in the vein grafts. In a smaller subset of patients (CLAS-II) who continued for 2 more years, significantly fewer patients on the double-drug regimen developed new lesions in native coronaries and in the bypass grafts. These results indicated the value of early initiation of vigorous lipid-lowering therapy in patients undergoing CABG. Multivariate analysis of factors which might explain these results showed that the predominant risk factor predicting global progression in the placebo group was non-HDL cholesterol. The predominant factor in the drug-treated subjects was apo C-III in HDL; this pointed to the importance of triglyceride-rich lipoproteins. This result was also seen in the MARS trial referred to earlier.

The Familial Atherosclerosis Treatment Study (FATS) was a pioneering effort by Dr. Greg Brown and co-workers in Seattle. They looked at 146 men who were less than 62 years old, had increased apo B levels of >125 mg/dL, family histories of vascular disease, and coronary disease by angiography. Of this group, 120 completed the trial, which used quantitative coronary angiography to monitor the changes. Interestingly, the marked reduction in clinical events far outshadowed the small, but significant, improvements in the serial angiograms. This suggested that plaque stabilization or endothelial effects due to lipid lowering may have resulted in the marked clinical improvements seen. Also, baseline Lp(a) was a coronary risk factor. As noted elsewhere in this book, aggressive lowering of LDL-c without significant change in Lp(a) removes the ability of Lp(a) to predict events.

This suggests Lp(a) as a marker and not as a specific target of lipid-lowering therapy.

The Harvard Atherosclerosis Reversibility Project (HARP) is the only angiographic trial which did not report a beneficial effect of lipid lowering. The mean baseline LDL-c in this trial was 140 mg/dL and multidrug therapy resulted in an end-trial LDL-c of 86 mg/dL. Nonetheless, unlike LCAS, no significant change between groups was seen. It is interesting to speculate that the small size of the trial may have precluded finding significant results (the type II error). By comparison, the LCAS trial was 4 times as large.

The University of California, San Francisco, Specialized Center for Atherosclerosis Research (UCSF-SCOR) sought to determine if aggressive multidrug treatment of men and women with familial hypercholesterolemia was beneficial. The study group was young, with a mean age of 42 years, and without overt CHD. Similar to trials where LDL-c values were not as high, progression was seen in approximately 50% of the control group and in only 28% of the treatment group. Regression was significantly more likely in the treatment group. Concerns that lipid lowering might not be effective in women were dispelled by this trial, which showed a more impressive change among the women than the men. On-trial LDL-c correlated better than change from the baseline with percent area stenosis (Table 7.7).

Multiple Interventions

There are trials which used multiple interventions. These include the Stanford Risk Intervention Project (SCRIP), Heidelberg, and the Lifestyle Heart Trial. In these trials, a low-fat diet was coupled with regular exercise; and in some, with cigarette cessation, drugs and/or stress reduction as well. The diets in SCRIP and Heidelberg included 20% fat. In the Lifestyle

TABLE 7.7 — MULTIDRUG TRIALS

Study	*n*	Years	Gender	Drug Dosages	% Δ LDL	% Δ HDL	Comment
CLAS-I pts with CABG	162	2	Men	Niacin 4.3 g/d; colestipol 30 g/d	– 43	37	Regression seen in 16.2% of treated vs 2.4% of placebo group ($P = 0.002$)
CLAS-II	103	4	Men	Niacin 4.2 g/d; colestipol 30 g/d	– 40	37	Significantly more treated showed nonprogression and regression in native coronaries
FATS	120	2.5	Men	Conventional, niacin and colestipol; lovastatin and colestipol	(See subgroup values given below)	(See subgroup values given below)	Less progression, more regression, fewer clinical events in double-drug groups
	38	2.5	Men	Lovastatin 20 mg bid and colestipol 30 g/d	– 46	15	Clinical events: 3 of 46

FATS (cont.)	36	2.5	Men	Niacin 4 g/d and colestipol 30 g/d	– 32	43	Clinical events: 2 of 48
	46	2.5	Men	Colestipol in 46%	– 7	5	Clinical events: 10 of 52
HARP	79	2.5	Men and women	Pravastatin + niacin: 38 pts; CME: 24 pts; gemfibrozil: 112 pts	– 41	13	No significant change in angiographic appearance between groups; end-trial LDL-c = 86 mg/dL
UCSF-SCOR	72	2	Men and women	Colestipol + niacin + lovastatin	– 39	26	Change in % area stenosis correlated with on-trial LDL-c

Abbreviations: *n*, number of patients; LDL, low-density lipoprotein; HDL, high-density lipoprotein; CLAS, Cholesterol Lowering Atherosclerosis Study; CABG, coronary artery bypass graft; FATS, Familial Atherosclerosis Treatment Study; HARP, Harvard Atherosclerosis Reversibility Project; CME, cholestyramine; pts, patients; LDL-c, low-density lipoprotein cholesterol; UCSF-SCOR, University of California, San Francisco, Specialized Center for Atherosclerosis Research.

Heart Trial, the diet was vegetarian with less than 10% fat. In the Heidelberg trial, intense physical activity was correlated with regression, whereas low levels of activity were not. The Lifestyle Heart Trial had the most dramatic improvements. This trial suffered from problems with randomization, and it was not easy to generalize the results seen in this small trial with those of additional large-scale trials. In SCRIP, multiple regression analysis identified the intake of dietary fat consumed during the study as the best correlate of new lesion formation. Participants randomized to the risk-reduction group preferentially increased complex carbohydrate intake to offset the reduction in dietary fat restriction. The authors suggested that the reduction in dietary fat decreased the rate of new lesion formation by mechanisms not limited to LDL-c reduction (Table 7.8).

Large-Scale Clinical Trials

■ Secondary Prevention Trials
(Tables 7.9 and 7.10)

The landmark Scandinavian Simvastatin Survival Study (4S) convincingly showed that intensive lipid lowering in survivors of MI could not only reduce rates of fatal and nonfatal CHD, but also lower total mortality as well. In this trial, average before-trial LDL-c was 190 mg/dL; after treatment, it was lowered to 130 mg/dL. Importantly, the large-scale nature of this trial showed that the benefits extended to older as well as younger subjects. Women also had reductions in cardiovascular end points. When the cost of reduced hospital admissions was taken into account, the resultant reduction in hospital costs over the 5.4 years of the trial reduced the effective cost of simvastatin by 88% to 28¢ per day.

The Cholesterol and Recurrent Events Trial (CARE) enrolled patients with lower levels of cho-

TABLE 7.8 — DIET/EXERCISE TRIALS

Study	*n*	Years	Gender	Interventions	% Δ LDL	% Δ HDL	Comment
Heidelberg	113	1	Men	AHA phase 3 diet with <20% fat, <200 mg cholesterol; exercise regimen	– 9	2	Less progression in the diet/exercise group than controls (23% vs 48%) and more regression (32% vs 17%)
Lifestyle	28	1	Men and women	Very low fat (<10%) vegetarian diet, exercise, stress management; smoking cessation	– 37	– 3	Progression in 53% of controls, 18% of treatment group; regression in 82% of treatment group
SCRIP	200	4	Men and women	Diet, exercise, weight reduction, smoking cessation, lipid medications	– 23	12	Regression seen in 10% of controls and in 21% of the treatment group; new lesion formation significantly reduced in treatment group

Abbreviations: *n*, number of patients; LDL, low-density lipoprotein, HDL, high-density lipoprotein; AHA, American Heart Association; SCRIP, Stanford Risk Intervention Project.

TABLE 7.9 — LARGE-SCALE SECONDARY PREVENTION TRIALS

Study	*n*	Years	Gender	Drug Dosages	% Δ LDL	% Δ HDL	Comment
4S	4444*	5.4	Men and women with previous MI	Simvastatin 20 mg; 37% took 40 mg	– 35	8	Relative risk of death reduced 30%; 42% reduction in coronary deaths; 37% reduction in revascularization procedures
CARE	4159†	5	Men and women with previous MI	Pravastatin 40 mg/d	– 28	2	CHD death and nonfatal MI were reduced 24%; 26% reduction in CABG; 22% reduction in PTCA
CDP (niacin arm)	1119	15	Men with previous MI	Niacin 3.0 g/d	– 10.1 for Δ serum cholesterol at 1 y	Not available	Patients with largest decrease in serum cholesterol at 1 y had lowest mortality; mortality in niacin group 11% lower than in placebo group

LIPID	9014	6	Men and women with preexisting CHD (31 to 75 y)	Pravastatin 40 mg/d	– 25	6	CHD death was reduced by 24%; total mortality was reduced 23% ($P = 0.00002$); fatal CHD and nonfatal MI were decreased 23%

Abbreviations: *n*, number of patients; LDL, low-density lipoprotein; HDL, high-density lipoprotein; 4S, Scandinavian Simvastatin Survival Study; MI, myocardial infarction; CARE, Cholesterol and Recurrent Events Trial; CHD, coronary heart disease; CABG, coronary artery bypass graft; PTCA, percutaneous transluminal coronary angioplasty; CDP, Coronary Drug Project; LIPID, Long-Term Intervention with Pravastatin in Ischemic Disease.

* Total cholesterol was 213 to 310 mg/dL; mean LDL-c, 190 mg/dL.

† Total cholesterol below 240 mg/dL with LDL-c-115 to 174 mg/dL; mean LDL-c, 139 mg/dL.

TABLE 7.10 — POST-CABG TRIALS

Study	*n*	Years	Gender	Drug Dosages	% Δ LDL	% Δ HDL	Comment
Post-CABG (aggressive treatment arm)	676	4.3	Men and women	Mean dose of 76 mg/d lovastatin; 30% of patients also given 8 g/d of cholestyra-mine	– 37 to – 40	7.5	In the aggressively treated group, about 66% of patients had an LDL-c below 100 mg/dL; mean % of grafts with atherosclerotic progression was 27% vs 39% in the moderately treated arm; rate of revascularization was 29% lower than in the moderately treated group
Post-CABG (moderate treatment arm)	675	4.3	Men and women	Mean dose of 4 mg/d lovastatin; 5% of patients also given 8 g/d cholestyramine	– 13 to – 15	5	In the moderately treated group, about 5% of patients had LDL-c below 100 mg/dL and 58% had a level of 130 mg/dL or higher

Abbreviations: *n*, number of patients; LDL, low-density lipoprotein; HDL, high-density lipoprotein; CABG, Coronary Artery Bypass Graft Trial; LDL-c, low-density lipoprotein cholesterol.

lesterol and extended the age cutoff to 75 years. CARE confirmed the 4S results with a highly significant reduction in the primary end point of CHD death and nonfatal MI. Analysis of subgroups showed that women and the elderly benefited as well. Lipid reduction in women had quite a pronounced effect, as they enjoyed a 45% reduction in the primary endpoint as compared with their male counterparts.

The Long-Term Intervention with Pravastatin in Ischemic Disease (LIPID) trial was the largest of the double-blind, randomized, placebo-controlled trials of secondary prevention. This trial enrolled 9014 men and women aged 31 to 75 years. The average baseline cholesterol levels in these patients with documented CHD (either acute MI or hospitalization for unstable angina pectoris) were 155 to 271 mg/dL. These values are intermediate to CARE and 4S. This trial also enrolled the largest number of women with CHD (1516 or 17% of participants) of any of the large-scale trials. This was particularly important because the finding of an increased risk of breast cancer seen in CARE was not seen in LIPID. Like 4S, there was a highly significant reduction of 24% not only in the primary end point, CHD mortality ($P = 0.00002$), but total mortality (23%) as well ($P = 0.00002$). There was a significant reduction in fatal and nonfatal CHD. Finally, as in CARE and 4S, there was a significant reduction in stroke by 20% ($P = 0.022$). To reduce the findings to absolute terms, the number needed to treat (NNT) with pravastatin therapy to prevent one fatal or important nonfatal cardiovascular event (MI or stroke) was 20 patients treated over the 6-year course of the trial.

The Coronary Drug Project (CDP) looked at both niacin (nicotinic acid) and clofibrate in survivors of MI. Neither drug showed a significant reduction in total mortality (the primary end point of the study) during the period of active drug treatment. Lipoproteins were not measured. Nonetheless, those who received niacin and showed the largest decrease in serum cholesterol at 1 year experienced a lower subsequent mortality than did those whose serum cholesterol values increased. Importantly, at a mean follow-up of 15 years, mortality from all causes was 11% lower in those assigned to 3 g of niacin daily than in the placebo group. This was a late benefit of a drug which in most cases had been discontinued earlier.

The Post Coronary Artery Bypass Graft Trial (Post-CABG) examined the benefit of aggressive LDL-c lowering in patients with a history of CABG (Table 7.10). The investigators randomized men (1249) and women (102) into either an aggressive treatment group receiving lovastatin (and in some cases, additional cholestyramine resin) to keep the mean LDL-c under 100 mg/dL or to a moderate treatment group given a smaller amount of lipid-lowering therapy to keep the LDL-c in the 132 to 136 mg/dL range. The trial used repeat angiography at a mean duration of 4.3 years to determine the primary end point, which was the percentage of initially patent major grafts per patient that showed substantial progression of atherosclerosis at follow-up. Those assigned to the aggressive treatment group had only a 27% rate of progression as contrasted with a rate of 39% for those who received moderate treatment ($P < 0.001$). Moreover, the aggressively treated group also had a lower rate of revascularization at 4 years of 6.5% as contrasted with a rate of 9.2% in those with the moderate lowering of LDL-c. The study also looked at low-dose warfarin in a 2-by-2 factorial design. Low-dose warfarin resulted in an international normalized ratio of only 1.4, which did not reduce the atherosclerotic progression in the CABGs. This study validates the effectiveness of statin therapy in achieving the National Cholesterol Education Program goal of an

LDL-c level of less than 100 mg/dL for those with coronary disease.

■ Primary Prevention Trials

Although the ground-breaking Lipid Research Clinics Coronary Primary Prevention Trial (LRC-CPPT) and the Helsinki Heart Trial (HHT) showed that lipid lowering could reduce the risk of a first MI, they were not sufficiently powered to show whether or not total mortality was affected by lipid lowering. The West of Scotland Coronary Prevention Study (WOSCOPS) was finally able to answer the question as to whether aggressive lipid lowering in high-risk men without a prior MI was safe. The improvement in overall mortality by 22%, which almost reached statistical significance ($P = 0.0051$), indicated that pravastatin therapy was indeed safe. It should be noted that these were high-risk men who had risk factors for CHD and an average age of approximately 55 years (Table 7.11). A further analysis showed that within this trial, those at the highest risk as determined by assessment of risk factors gained the most from pravastatin therapy.

The Air Force/Texas Coronary Atherosclerosis Prevention Study (AFCAPS/TexCAPS) extended these observations by showing that middle-aged men and women reduced significantly the first coronary event with lovastatin therapy. Criteria for entry included a total cholesterol of 180 to 264 mg/dL, a raised LDL-c (130 to 190 mg/dL), and an HDL-c ≤45 for men or ≤47 for women. This study also included as a primary end point the combined determination of fatal and nonfatal MI as well as unstable angina. The authors utilized unstable angina as the primary end point (as in the FATS trial) to be more representative of the growing trend in acute coronary syndromes. In the past decade, the number of nonfatal MIs has been decreasing and the admissions for unstable angina increasing.

TABLE 7.11 — LARGE-SCALE PRIMARY PREVENTION TRIALS							
Study	***n***	**Years**	**Gender**	**Drug Dosages**	**% Δ LDL**	**% Δ HDL**	**Comment**
AFCAPS/ TexCAPS	6605	4.8	85% men 45 to 73 years at entry; 15% women 55 to 73 years at entry	Lovastatin 20 mg or 40 mg titrated to reduce LDL-c to <110; average dose 30 mg/d	– 25	+ 6	36% reduction in first coronary event; 26% reduction in fatal and nonfatal MI; 33% reduction in revascularization (CABG, PTCA)
HHT	4081	5	Men	Gemfibrozil 600 mg bid	– 10	10	34% reduction in incidence of CHD (CI, 8.2 to 52.6)
LRC-CPPT	3806	7.4	Men	Cholestyramine resin 24 g/d	– 20.3	1.6	19% reduction in risk of fatal and nonfatal MI; no decrease in total mortality

WOSCOPS	6595	4.9	High-risk men; mean age 55 years	Pravastatin 40 mg/d	– 26	5	31% reduction in fatal and nonfatal MI; 32% reduction from all cardiovascular causes; 22% reduction in all-cause mortality ($P = -0.051$)

Abbreviations: *n*, number of patients; LDL, low-density lipoprotein; HDL, high-density lipoprotein; AFCAPS/TexCAPs, Air Force/Texas Coronary Atherosclerosis Prevention Study; MI, myocardial infarction; CABG, coronary artery bypass graft; PTCA, percutaneous transluminal coronary angioplasty; HHT, Helsinki Heart Trial; CHD, coronary heart disease; CI, confidence interval; LRC-CPPT, Lipid Research Clinics Coronary Primary Prevention Trial; WOSCOPS, West of Scotland Coronary Prevention Study.

The fact that two large-scale studies have been strikingly positive suggests that statin therapy with drugs such as lovastatin and pravastatin may indeed be indicated to prevent a first coronary event in those at the same level of risk as the participants in these trials. Cohort studies such as the Framingham have emphasized the importance of age in the risk equation. In both WOSCOPS and AFCAPS/TexCAPS, the average age of the participants was in the mid 50s to assure that these were indeed high-risk patients. These data do not argue for starting statin therapy in premenopausal women or in men under 35 who have LDL-c under 220 mg/dL and whose risk-factor profile is otherwise low.

Taken together, these clinical trials provide valuable information that can be used to decide who should be treated with statins to prevent a first coronary event. Data from Framingham compiled into a "risk score" help determine an individual's absolute risk. This gives a useful overview of an individual's risk, which can be brought into sharper focus by emerging noninvasive studies such as electron-beam computed tomography (EBCT) or carotid imaging. While these studies should not be used to "screen" the general population, they may have important utility in helping decide who among those with risk factors actually has evidence for early subclinical disease. The advisability of drug treatment, however, is gleaned from clinical trial data. Table 7.12 shows how data from two primary- and two secondary-prevention trials can be combined to provide the NNT, a useful summary statistic (Cook and Sackett, 1995). Other derived statistics include years of life gained, adverse events sustained to prevent one outcome, and cost analyses (Robson, 1997). How this information is presented may greatly influence clinical decisions. The point is not a trivial one; the accurate determination of clinical treatment thresholds has important economic implications.

■ Apheresis Trials

Apheresis is a superb way to lower raised LDL-c and also lower raised concentrations of Lp(a). Its major disadvantages have been time and cost. To examine its benefits/risks more closely, the LDL Apheresis Atherosclerosis Regression Study (LAARS) looked at 42 men with severe hypercholesterolemia who had LDL-c >225 mg/dL, triglycerides <440 mg/dL, and angiographic CHD (Table 7.13). Half were randomized to receive simvastatin 40 mg/d and half received simvastatin as well as biweekly LDL apheresis. There were no statistically significant differences in mean lumen diameters or minimum lumen diameters between the two groups. The apheresis patients did show a significant improvement in regional myocardial perfusion measured by hyperemic mean transit time in contrast to no change in the patients treated with simvastatin alone. Another trial of familial hypercholesterolemia subjects utilizing quantitative coronary angiography at 2 years also failed to show an advantage to the apheresis group as contrasted with the drug group even though favorable changes in Lp(a) as well as LDL-c were seen (Thompson et al, 1995). The small sample sizes involved may have precluded seeing a small change that would have been more evident with larger trials and longer durations.

■ Non–LDL-Lowering Trials

The trials referred to at this point in the chapter have all had as their primary focus the lowering of LDL-c. Two recent clinical trials have utilized therapies to improve dyslipidemia (eg, high triglycerides and/or low HDL-c) without substantially changing levels of LDL-c (Table 7.14). Although the mechanisms for benefit have yet to be elucidated, this does

TABLE 7.12 — NUMBERS NEEDED TO TREAT, CALCULATED FROM LARGE-SCALE PREVENTION TRIALS TO PREVENT ONE END POINT EVENT

Name of Trial	Trial Notes	Number Needed to Treat
4S	Secondary prevention; Rx: Simvastatin 20 or 40 mg/d; mean baseline LDL = 189	For all CAD deaths and major CAD events = 10
CARE	Secondary prevention; Rx: Pravastatin 40 mg/d; mean baseline LDL = 139	For all fatal and nonfatal events = 33
WOSCOPS	Primary prevention; Rx: Pravastatin 40 mg/d; mean baseline LDL = 192	For all fatal and nonfatal events = 48
AFCAPS/TexCAPS	Primary prevention; Rx: Lovastatin 40 mg/d; mean baseline LDL = 156	For risk of first major coronary event = 58

Abbreviations: 4S, Simvastatin Scandinavian Trial; Rx, therapy; LDL, low density lipoiprotein; CAD, cornary artery disease; CARE, Cholesterol and Recurrent Events Trial; WOSCOPS, West of Scotland Coronary Prevention Study; AFCAPS/TexCAPS, Air Force/Texas Coronary Atherosclerosis Prevention Study.

TABLE 7.13 — LOW-DENSITY LIPOPROTEIN APHERESIS TRIAL

Study	*n*	Years	Gender	Drug Dosages	% Δ LDL	% Δ Lp(a)	Comment
FHRS	20	2.1	Men with FH	Apheresis + simvastatin 40 mg/d	—	Reduced	Apheresis lowered LDL-c and Lp(a) more but didn't cause improvement in coronary angiographic parameters
FHRS	19	2.1	Men with FH	Simvastatin 40 mg/d + colestipol 20 g/d	—	—	—
LAARS	21	2	Men with FH	Apheresis biweekly and simvastatin 40 mg/d	63	– 19	No statistically significant increase in mean lumen diameters over simvastatin alone; did show improvement in regional myocardial perfusion
LAARS	21	2	Men with FH	Simvastatin 40 mg/d	47	+ 15	—

Abbreviations: *n*, number of patients; LDL, low-density lipoprotein; HDL, high-densitiy lipoprotein; FHRS, Familial Hypercholesterolemia Regression Study; FH, familial hypercholesterolemia; LAARS, LDL Apheresis Atherosclerosis Regression Study.

7

TABLE 7.14 — NON-LDL-C-LOWERING SECONDARY PREVENTION TRIALS

Study	*n*	Years	Gender	Drug*	% Δ LDL	% Δ HDL	Comment
BECAIT	92	5	Men; <45 y of age	Bezafibrate 200 mg tid	0	+ 9	TGs lowered 31%; significant change in angiographic and clinical end points
LOCAT	395	1.3	Men; <70 y of age and post-CABG	Gemfibrozil 1200 mg/d	– 11	0 (in trial)	Improvement in angiographic end points; fewer new lesions in bypass grafts; TG-rich lipoproteins correlated with progression
VA HIT	2531	5.1	Men with CHD	Gemfibrozil 1200 mg/d	0	+ 6 (TG: – 31)	Absolute risk reduction 4.4%; relative risk reduction 22%; significant 24% reduction in combined outcome CHD death, nonfatal MI, and stroke

Abbreviations: *n*, number of patients; LDL, low-density lipoprotein; HDL, high-density lipoprotein; BECAIT, Bezafibrate Coronary Atherosclerosis Intervention Trial; TG, triglyceride; LOCAT, Lopid Coronary Angiography Trial; CABG, coronary artery bypass graft; VA HIT, Veterans Affairs High-Density Lipoprotein Cholesterol Intervention Trial; CHD, coronary heart disease; MI, myocardial infarction.

*Intervention drug

not deny the key role of LDL in atherosclerosis. For example, the improvements in HDL-c and triglycerides may make LDL less susceptible to oxidation. Nonetheless, data from the Monitored Atherosclerosis Regression Study (MARS), the Cholesterol Lowering Atherosclerotic Study (CLAS), and the NHLBI Type II Study indicate that triglyceride-rich lipoproteins may also be of importance for the progression of mild/moderate lesions in subjects treated with cholesterol-lowering regimens. They argue that therapy too narrowly focused on LDL-c alone may not provide optimum therapy for all patients at risk or with CHD.

The Bezafibrate Coronary Atherosclerosis Intervention Trial (BECAIT) was a 5-year, placebo-controlled, randomized trial that recruited 92 men following an MI. Initial cholesterol and triglycerides had to be under 200 and 150 mg/dL, respectively, at baseline. Bezafibrate, a fibric acid derivative, decreased total serum cholesterol by 9%, triglycerides by 31%, and increased HDL-c by 9% without a significant change in LDL-c. There was a significant improvement in the mean minimum luminal diameter on serial angiography. Of note, there was a significant decrease in clinical coronary events in the treated group as compared with the placebo group ($P = 0.02$ log rank test).

The Lopid Coronary Angiography Trial (LOCAT) used a sustained release 1200-mg tablet in nondiabetic, nonsmoking, nonobese men with CHD who had low HDL-c and high triglycerides. There was a significant improvement in the change in per-patient means of average diameters of coronary segments as well as in the minimal diameters of stenotic lesions. In venous bypass grafts, more subjects on placebo (14%) had new lesions on serial angiography than those assigned to gemfibrozil (Lopid) (2%) ($P = 0.001$). Triglyceride-rich lipoproteins predicted progression in this study.

7

The Veterans Affairs High-Density Lipoprotein Cholesterol Intervention Trial (VA HIT) was a randomized clinical trial comparing gemfibrozil (1200 mg/d) with placebo in 2531 men with CHD. The metabolic profile was different from the statin trials. This group had mean HDL-c levels of 32 mg/dL; mean LDL-c levels of 111 mg/dL; mean levels of triglycerides of 160 mg/dL; and a mean age of 64 years. There was a higher prevalence of abnormalities associated with the metabolic syndrome than seen in the statin trials. The prevalence of diabetes was 25%; hypertension, 57%; and a mean waist-to-hip ratio of 0.96 indicative of abdominal obesity. Without lowering LDL-c levels, the therapy caused an absolute risk reduction of 4.4% with a reduction in relative risk of 22% (95% confidence interval, 7% to 35%; $P = 0.006$). There was a 24% reduction in the combined outcome of death from CHD, nonfatal MI, and stroke ($P < 0.001$).

■ Carotid Artery Disease and Lipid-Lowering Trials

Early clinical trials did not indicate that lowering cholesterol resulted in a reduction of stroke incidence. Yet improvements seen in stroke incidence in 4S and CARE suggested that with significant lowering of LDL-c in those with clinical coronary atherosclerosis there is an important impact on the end point of stroke. An analysis of secondary prevention trials showed that use of statins resulted in a reduction of stroke by 32% ($P = 0.001$) (Crouse et al, 1997). Insight into the potential mechanism of benefit of LDL-c lowering has come from the measurement of IMT of the carotid artery (multiple sites of measurement) in numerous studies (Table 7.15). Using either statin therapy or apheresis, these studies showed improvements in IMT as compared with placebo. In the MARS, when the major apolipoprotein B–containing lipoproteins were measured independently, intermediate-density lipoprotein, but not VLDL or LDL, was associated with progression of carotid artery IMT (Hodis et al, 1997). Patients with carotid bruits and evidence of carotid atherosclerotic disease would appear to benefit from statin therapy. For these patients at high risk for subclinical coronary disease, an argument can be made for an LDL-c goal of 100 mg/dL or less.

> ***Clinical Tip:*** Noninvasive imaging of the carotid arteries is a potential tool for following the results of lipid-lowering therapy. In MARS, for every 0.03-mm increase per year in carotid IMT there was a significant increase in the relative risk for both fatal and nonfatal MI and for any coronary event (Hodis et al, 1998). It requires, however, a skilled operator in carotid ultrasonographic arterial imaging. The clinical trials listed in Table 7.15 provide further details.

TABLE 7.15 — CAROTID DISEASE AND LDL–C-LOWERING INTERVENTION TRIALS

Study	*n*	Years	Gender	Drug Dosages	% Δ LDL	% Δ HDL	Comment
ACAPS	919	3	Men and women	Lovastatin 26 mg/d	– 28	5	Lovastatin significantly reversed progression of IMT; fewer morbid and fatal CV events with lovastatin
CAIUS	305	3	Men and women	Pravastatin 40 mg	– 23	3	Progression of IMT significantly affected by pravastatin
KAPS	447	3	Men	Pravastatin 40 mg/d	– 27.4	– 1.9	IMT improved by 45% with pravastatin
LAARS	21	2	Men with FH	Apheresis biweekly and simvastatin 40 mg/d	63	– 19	Apheresis + simvastatin signficantly reversed progression of IMT seen with simvastatin alone, whereas no change in coronary disease

MARS	188 at 2 y; 74 at 4 y	4	Men; all <45 y of age	Lovastatin 80 mg	—	—	Regression with lovastatin and progression with placebo; annual rate of change in IMT significantly correlated with on-trial concentrations of LDL-c, TG, apo B, apo C III, and apo E

Abbreviations: *n*, number of patients; LDL, low-density lipoprotein; HDL, high-density lipoprotein; ACAPS, Asymptomatic Carotid Artery Progression Study; IMT, intimal medial thickness; CV, cardiovascular; CAIUS, Carotid Atherosclerosis Italian Ultrasound Study; KAPS, Kuopio Atherosclerosis Prevention Study; LAARS, LDL Apheresis Atherosclerosis Regression Study; FH, familial hypercholesterolemia; MARS, Monitored Atherosclerosis Regression Study; LDL-c, low-density lipoprotein cholesterol; TG, triglyceride; apo, apolipoprotein.

7

SUGGESTED READINGS

ACAPS

Furberg CD, Adams HP Jr, Applegate WB, et al. Effect of lovastatin on early carotid atherosclerosis and cardiovascular events. Asymptomatic Carotid Artery Progression Study (ACAPS) Research Group. *Circulation.* 1994;90:1679-1687.

AFCAPS/TexCAPS

Downs JR, Beere PA, Whitney E, et al. Design & rationale of the Air Force/Texas Coronary Atherosclerosis Prevention Study (AFCAPS/TexCAPS). *Am J Cardiol.* 1997;80:287-293.

BECAIT

Ericsson CG, Hamsten A, Nilsson J, Grip L, Svane B, de Faire U. Angiographic assessment of effects of bezafibrate on progression of coronary artery disease in young male postinfarction patients. *Lancet.* 1996;347:849-853.

Ericsson CG, Nilsson J, Grip L, Svane B, Hamsten A. Effect of bezafibrate treatment over five years on coronary plaques causing 20% to 50% diameter narrowing (The Bezafibrate Coronary Atherosclerosis Intervention Trial [BECAIT]). *Am J Cardiol.* 1997;80:1125-1129.

Cardioprotective Diet Study from India

Singh RM, Rastogi SS, Verma R, et al. Randomised controlled trial of cardioprotective diet in patients with recent acute myocardial infarction: results of one year follow up. *BMJ.* 1992;304:1015-1019.

CAIUS

Mercuri M, Bond MG, Sirtori CR, et al. Pravastatin reduces carotid intima-media thickness progression in an asymptomatic hypercholesterolemic mediterranean population: the Carotid Atherosclerosis Italian Ultrasound Study. *Am J Med.* 1996;101:627-634.

CARE

Sacks FM, Pfeffer MA, Moye LA, et al. The effect of pravastatin on coronary events after myocardial infarction in patients with average cholesterol levels. Cholesterol and Recurrent Events Trial investigators. *N Engl J Med.* 1996;335:1001-1009.

CCAIT

Waters D, Higginson L, Gladstone P, et al. Effect of monotherapy with an HMG-CoA reductase inhibitor on the progression of coronary atherosclerosis as assessed by serial quantitative angiography. The Canadian Coronary Atherosclerosis Intervention Trial. *Circulation.* 1994;89:959-968.

CDP

Canner PL, Berge KG, Wenger NK, et al. Fifteen year mortality in Coronary Drug Project patients: long-term benefit with niacin. *J Am Coll Cardiol.* 1986;8:1245-1255.

CLAS I

Blankenhorn DH, Nessim SA, Johnson RL, Sanmarco ME, Azen SP, Cashin-Hemphill L. Beneficial effects of combined colestipol-niacin therapy on coronary atherosclerosis and coronary venous bypass grafts [published erratum appears in *JAMA.* 1988;259:2698]. *JAMA.* 1987;257:3233-3240.

DART Trial

Burr ML, Fehily AM, Gilbert JF, et al. Effects of changes in fat, fish and fibre intakes on death and myocardial infarction: diet and reinfarction trial (DART). *Lancet.* 1989;2:757-761.

FATS

Brown G, Albers JJ, Fisher LD, et al. Regression of coronary artery disease as a result of interim lipid-lowering therapy in men with high levels of apolipoprotein B. *N Engl J Med.* 1990;323:1289-1298.

FHRS

Thompson GR, Maher VM, Matthews S, et al. Familial Hypercholesterolemia Regression Study: a randomised trial of low-density-lipoprotein apheresis. *Lancet* 1995;345:811-816.

HARP

Sacks FM, Pasternak RC, Gibson CM, Rosner B, Stone PH. Effect on coronary atherosclerosis of decrease in plasma cholesterol concentrations in normocholesterolaemic patients. Harvard Atherosclerosis Reversibility Project (HARP) Group. *Lancet.* 1994;344:1182-1186.

Heidelberg Study

Schuler G, Hambrecht R, Schlierf G, et al. Regular physical exercise and low-fat diet. Effects of progression on coronary artery disease. *Circulation.* 1992;86:1-11.

KAPS

Salonen R, Nyyssönen K, Porkkala E, et al. Kuopio Atherosclerosis Prevention Study (KAPS). A population-based primary preventive trial of the effect of LDL lowering on atherosclerotic progression in carotid and femoral arteries. *Circulation.* 1995;92:1758-1764.

LAARS

Aengevaeren WR, Kroon AA, Stalenhoef AF, Uijen GJ, van der Werf T. Low density lipoprotein apheresis improves regional myocardial perfusion in patient with hypercholesterolemia and extensive coronary artery disease. The LDL-Apheresis Atherosclerosis Regression Study (LAARS). *J Am Coll Cardiol.* 1996;28:1696-1704.

LCAS

Herd JA, Ballantyne CM, Farmer JA, et al. Effects of fluvastatin on coronary atherosclerosis in patients with mild to moderate cholesterol elevations. (Lipoprotein and Coronary Atherosclerosis Study [LCAS]). *Am J Cardiol.* 1997;80:278-286.

West MS, Herd JA, Ballantyne CM, et al. The Lipoprotein and Coronary Atherosclerosis Study (LCAS): design, methods, and baseline data of a trial of fluvastatin in patients without severe hypercholesterolemia. *Control Clin Trials.* 1996;17:550-583.

Lifestyle Heart Trial

Ornish D, Brown SE, Scherwitz LW, et al. Can lifestyle changes reverse coronary heart disease? The Lifestyle Heart Trial. *Lancet.* 1990;336:129-133.

LIPID

The Lipid Study Group. Design features and baseline characterisitcs of the LIPID (Long-Term Intervention with Pravastatin in Ischemic Disease) Study: a randomized trial in patients with previous acute myocardial infarction and/or unstable angina pectoris. *Am J Cardiol.* 1995;76:474-479.

LOCAT

Frick MH, Syvanne M, Nieminen MS, et al. Prevention of the angiographic progression of coronary and vein-graft atherosclerosis by gemfibrozil after coronary bypass surgery in men with low levels of HDL cholesterol. Lopid Coronary Angiography Trial (LOCAT) Study Group. *Circulation.* 1997;96:2137-2143.

LRC-CPPT

The Lipid Research Clinics Coronary Primary Prevention Trial results. I. Reduction in incidence of coronary heart disease. *JAMA*. 1984;251:351-364.

MAAS

MAAS investigators. Effect of simvastatin on coronary atheroma: the Multicentre Anti-Atheroma Study (MAAS) [published erratum appears in *Lancet*. 1994;344:762]. *Lancet*. 1994;344:633-638.

MARS

Blankenhorn DH, Azen SP, Kramsch DM, et al. Coronary angiographic changes with lovastatin therapy. The Monitored Atherosclerosis Regression Study (MARS). The MARS Research Group. *Ann Intern Med*. 1993;119:969-976.

Mack WJ, Krauss RM, Hodis HN. Lipoprotein subclasses in the Monitored Atherosclerosis Regression Study (MARS). Treatment effects and relation to coronary angiographic progression. *Arterioscler Thromb Vasc Biol*. 1996;16:697-704.

Shircore AM, Mack WJ, Selzer RH, et al. Compensatory vascular changes of remote coronary segments in response to lesion progression as observed by sequential angiography from a controlled clinical trial. *Circulation*. 1995;92:2411-2418.

Mediterranean Diet Study

de Lorgeril M, Renaud S, Mamelle N, et al. Mediterranean alpha-linolenic acid-rich diet in secondary prevention of coronary heart disease [published erratum appears in *Lancet*. 1995;345: 738]. *Lancet*. 1994;343:1454-1459.

NHLBI Type II Intervention Study

Brensike JF, Levy RI, Kelsey SF, et al. Effects of therapy with cholestyramine on progression of coronary arteriosclerosis: results of the NHLBI Type II Coronary Intervention Study. *Circulation*. 1984;69:313-324.

PEPI Trial

The Writing Group for the PEPI Trial. Effects of estrogen or estrogen/progestin regimens on heart disease risk factors in postmenopausal women. The Postmenopausal Estrogen/Progestin Intervention (PEPI) Trial [published erratum appears in *JAMA*. 1995;274:1676]. *JAMA*. 1995;273:199-208.

PLAC I

Pitt B, Mancini GB, Ellis SG, Rosman HS, Park JS, McGovern ME. Pravastatin limitation of atherosclerosis in the coronary arteries (PLAC I): reduction in atherosclerosis progression and clinical events. PLAC-I investigation. *J Am Coll Cardiol.* 1995;26:1133-1139.

PLAC II

Crouse JR 3rd, Byington RP, Bond MG, et al. Pravastatin, Lipids, and Atherosclerosis in the Carotid Arteries (PLAC-II) [published erratum appears in *Am J Cardiol.* 1995;75:862]. *Am J Cardiol.* 1995;75: 455-459.

POSCH Trial

Buchwald H, Varco RL, Matts JP, et al. Effect of partial ileal bypass surgery on mortality and morbidity from coronary heart disease in patients with hypercholesterolemia. Report of the Program on the Surgical Control of the Hyperlipidemias (POSCH). *N Engl J Med.* 1990;323:946-955.

Post-CABG Trial

The Post Coronary Artery Bypass Graft Trial Investigators. The effect of aggressive lowering of low-density lipoprotein cholesterol levels and low-dose anticoagulation on obstructive changes in saphenous-vein coronary artery bypass grafts [published erratum appears in *N Engl J Med.* 1997;377;1859]. *N Engl J Med.* 1997;336:153-162.

Pravastatin Multinational Study Group for Cardiac Risk Patients

The Pravastatin Multinational Study Group for Cardiac Risk Patients. Effects of pravastatin in patients with serum total cholesterol levels from 5.2 to 7.8 mmol/liter (200-300 mg/dl) plus two additional atherosclerotic risk factors. *Am J Cardiol.* 1993; 72:1031-1037.

REGRESS

Jukema JW, Bruschke AV, van Boven AJ, et al. Effects of lipid lowering by pravastatin on progression and regression of coronary artery disease in symptomatic men with normal to moderately elevated serum cholesterol levels. The Regression Growth Evaluation Statin Study. *Circulation.* 1995;91:2528-2540.

SCRIP

Quinn TG, Alderman EL, McMillan A, Haskell W. Development of new coronary atherosclerotic lesions during a 4-year multifactor risk reduction program: the Stanford Coronary Risk Intervention Project (SCRIP). *J Am Coll Cardiol.* 1994;24:900-908.

Scandinavian Simvastatin Survival Study (4S)

Scandinavian Simvastatin Survival Study Group. Randomised trial of cholesterol lowering in 4444 patients with coronary heart disease: the Scandinavian Simvastatin Survival Study (4S). *Lancet.* 1994;344:1383-1389.

STARS

Watts GF, Lewis B, Brunt JN, et al. Effects on coronary artery disease of lipid-lowering diet, or diet plus cholestyramine, in the St Thomas' Atherosclerosis Regression Study (STARS). *Lancet.* 1992;339:563-569.

Watts GF, Mandalia S, Brunt JN, Slavin BM, Coltart DJ, Lewis B. Independent associations between plasma lipoprotein subfraction levels and the course of coronary artery disease in St Thomas' Atherosclerosis Regression Study (STARS). *Metabolism.* 1993; 42:1461-1467.

UCSF-SCOR

Kane JP, Malloy MJ, Ports TA, Phillips NR, Diehl JC, Havel RJ. Regression of coronary atherosclerosis during treatment of familial hypercholesterolemia with combined drug regimens. *JAMA.* 1990;264:3007-3012.

WOSCOPS

Shepherd J, Cobbe SM, Ford I, et al. Prevention of coronary heart disease with pravastatin in men with hypercholesterolemia. West of Scotland Coronary Prevention Study Group. *N Engl J Med.* 1995;333:1301-1307.

VA HIT

Rubins HB, Robins SJ, Collins D, et al. Gemfibrozil for the secondary prevention of coronary heart disease in men with low levels of high-density lipoprotein cholesterol. Veterans Affairs High-Density Lipoprotein Cholesterol Intervention Trial Study Group. *N Engl J Med.* 1999;341:410-418.

CAROTID ARTERY STUDIES

Crouse JR 3rd, Byington RP, Hoen HM, Furberg CD. Reductase inhibitor monotherapy and stroke prevention. *Arch Intern Med.* 1997;157:1305-1310.

Hodis HN, Mack WJ, Dunn M, et al. Intermediate-density lipoproteins and progression of carotid arterial wall intima-media thickness. *Circulation.* 1997;95:2022-2026.

Hodis HN, Mack WJ, La Bree L, et al. The role of carotid arterial intima-media thickness in predicting clinical coronary events. *Ann Intern Med.* 1998;128:262-269.

GENERAL

Ballantyne CM, Herd JA, Dunn JK, Jones PH, Farmer JA, Gotto AM Jr. Effects of lipid lowering therapy on progression of coronary and carotid artery disease. *Curr Opin Lipidol.* 1997;8:354-361.

Cook RJ, Sackett DL. The number needed to treat: a clinically useful measure of treatment effect [published erratum appears in *BMJ.* 1995;310:1056]. *BMJ.* 1995;310:452-454.

Robson J. Information needed to decide about cardiovascular treatment in primary care. *BMJ.* 1997;314:277-280.

8 Dietary Therapy for Hyperlipidemia

Diet is the hallmark of therapy for hyperlipidemia. Its usefulness extends beyond simply providing additional low-density lipoprotein cholesterol (LDL-c) lowering for those with hypercholesterolemia. It also can help improve the abnormal high-density lipoprotein cholesterol (HDL-c) and triglyceride values seen in combined hyperlipidemia. In fact, patients with combined hyperlipidemia are often quite sensitive to weight reduction and exercise. Diet may play an important role in those with low HDL-c and premature coronary heart disease (CHD). Knowledge of the effects of dietary fat, kind of dietary fat, effect of dietary carbohydrate, and alcohol may prove useful in the overall care of patients. In those with severe hypertriglyceridemia who are prone to pancreatitis, reduction of all dietary fats and cessation of alcohol is crucial to avoiding exacerbations that could lead to the chylomicronemia syndrome and acute pancreatitis. Also, dietary antioxidants may reduce risk of CHD through mechanisms independent of lipid effects. Finally, antithrombotic and antiarrhythmic effects of diet may prove useful.

The two diets recommended by the National Cholesterol Education Program are the Step I and Step II Diets (Table 8.1). The Step I diet is really the "population" diet. While some clinicians may contend that their patients will not change, it is important to note that over the past 35 years, the average intake of fat, saturated fat, and cholesterol has changed significantly and has been associated with important declines in serum cholesterol for both men and women. The sig-

TABLE 8.1 — MAIN FEATURES OF THE STEP I AND STEP II NCEP DIETS

Constituents of Diet	Step I Diet "Population"	Step II Diet "High Risk"
Total fat	*Goal:* 30% or less reduction measured as % of total energy	
Saturated fatty acids	8% to 10%	<7%
Polyunsaturated fat	Up to 10%	
Monounsaturated fat	Up to 15%	
Dietary cholesterol	<300 mg/d	<200 mg/d
Total calories	*Goal*: Number necessary to achieve and maintain desirable weight	

Abbreviation: NCEP, National Cholesterol Education Program.

nificance of this decline can be measured by considering the improvement in those with the highest levels of cholesterol as well as the improvement in the nation's mean blood cholesterol. Thus, according to National Health and Nutrition Examination Survey (NHANES) data, the proportion of adult men aged 20 through 74 with high blood cholesterol defined as 240 mg/dL or greater has decreased from 25% to 19%; and similarly, the proportion of adult women with high blood cholesterol has fallen from 28% to 20%. The age-adjusted mean serum cholesterol level for men and women over the past 30 years has fallen from a mean cholesterol level of 220 mg/dL in 1960 to 1962 to the current average of 205 mg/dL. More than half of the decrease occurred from 1976 to 1991.

The Step II diet is more rigorous and is designed to achieve greater restriction of dietary saturated fat. For a person to achieve less than 7% of calories from saturated fat and less than 200 mg/d of dietary cholesterol, the help of a registered dietitian is advised, as knowledge, skills, and support are needed to effect proper behavioral change. Diets very low in fat have been advocated by some physicians, but there are practical limitations that come into play when total fat is reduced below 15% of energy.

There are both biologic as well as clinical factors that determine the response of the individual to diet. Biologic factors include:

- Specific changes in dietary composition
- Initial serum cholesterol concentration
- Metabolic responsiveness
- Genetic background.

Clinical factors include:

- Compliance with diet
- Change in body weight.

Clinicians need to know and understand the changes in dietary composition that affect LDL-c, HDL-c, and triglycerides (Table 8.2). Denke has shown that the higher the initial serum cholesterol value, the greater the response to diet. Also, excess body weight is a factor that must be considered; it is a strong contributor to elevated cholesterol levels in men. Weight gain in adult life can lead not only to abnormal lipid values, but to increased risks for diabetes mellitus and CHD. Genetic influences on dietary responsiveness are considerable. Those patients with familial hypercholesterolemia cannot usually hope to achieve goal LDL-c values with diet alone. On the other hand, those with familial combined hyperlipidemia may improve markedly with diet, weight loss, and exercise. Those with apo E-IV phenotype absorb more dietary cholesterol than those with apo E-II. Also important are apo A-IV, genetic influences on the apo B and the apo A-I promoter region, and the LDL subclasses.

Dietary Influences on Lipids and Lipoproteins

The classic equation of Keys and Hegstedt established the important role of saturated fat and dietary cholesterol in determining change in blood cholesterol when changes were made in the metabolic ward setting. Their classic experiments showed that dietary saturated fat is twice as potent in raising blood cholesterol values as polyunsaturated fats are in lowering them. Dietary cholesterol is also important but quantitatively has less effect.

■ Dietary Cholesterol

Sources of dietary cholesterol include:

- Animal meat

- Egg yolk
- Organ meats such as liver.

Dietary cholesterol affects LDL-c by suppression of cholesterol production in the liver. While the effects of dietary cholesterol on some animal species is striking, with massive hypercholesterolemia seen, the effects in humans are highly variable. Hyper- and hyporesponders are seen. There are gender differences as well, with women increasing their HDL-c more than men with cholesterol feeding. Nonwhites appear to be particularly responsive to cholesterol feeding as compared with whites, in whom LDL-c increases more strikingly in response to saturated fat intake (Fielding et al, 1995).

■ Saturated Fats

Sources of saturated fat in the diet include:

- Fatty meats
- High-fat dairy products
- Vegetables sources such as coconut and palm kernel oil (Table 8.3).

Saturated fats appear to raise blood cholesterol and LDL-c through their influence on the LDL receptor. Among saturated fats, lauric, myristic, and palmitic are the most cholesterol raising (Table 8.3). Stearic acid has not been found to be cholesterol raising and may, in some circumstances, lower blood cholesterol. Nonetheless, in the St. Thomas Atherosclerosis Regression Study (STARS), progression of coronary disease was directly, strongly, and independently associated with intake of saturated fatty acids from lauric through stearic. This fact was not fully explained by the effects of these fatty acids on cholesterol. Certainly at the population level, dietary saturated fatty acids are strongly associated with CHD; this was best seen in the 16 cohorts from 14 countries followed in

TABLE 8.2 — DIETARY FACTORS THAT AFFECT LIPIDS/LIPOPROTEINS

	Lower LDL-c	Raise HDL-c	Lower Triglycerides
Fats			
Saturated	No, raise LDL-c	Yes	No
Monounsaturated	Yes*	No	Yes, if substitute for CHO
Polyunsaturated (n-6)	Yes*	No	Yes, if substitute for CHO
Trans fatty acids	No,† raise LDL-c	No†	No
Dietary cholesterol	No‡	Yes‡	No
Fiber	Yes	No	No
Soy protein	Yes§	No	No
Weight loss	Yes	Yes	Yes
Alcohol	No	Yes	No, raises triglycerides

Dietary carbohydrate	Yes*	Lowers HDL-c	No, raises triglycerides
Fish oil (n-3)	No[‖]	Variable	Yes

Abbreviations: LDL-c, low-density lipoprotein cholesterol; HDL-c, high-density lipoprotein cholesterol; CHO, carbohydrate.

* When substituted for saturated fat in the diet.
† In high amounts, raise LDL-c and lower HDL-c.
‡ Response to diet is variable; may increase LDL-c significantly in some but not in others; may increase HDL-c more in women than in men.
§ LDL-c lowering seen in hypercholesterolemia.
‖ May raise LDL-c as triglycerides are lowered in cases of combined hyperlipidemia.

8

TABLE 8.3 — SOURCES OF FATTY ACIDS IN THE DIET

Name	Structure*	Common sources
Saturated		
Lauric	C12:0	Coconut oil, palm kernel oil
Myristic	C14:0	Coconut oil
Palmitic	C16:0	Palm oil, beef
Stearic	C18:0	Cocoa butter, beef
Monounsaturated (n-9)†—oleic	C18:1	Olive oil, canola oil, beef, peanut oil, avocado
Polyunsaturated (n-6)†—linoleic	C18:2	Corn, cottonseed, safflower, sunflower, soybean oils
Polyunsaturated (n-3)†—linolenic	C18:3	Flaxseed oil, canola oil, soybean oil, nuts (English walnuts)
Eicosapentaenoic	C20:5	Fatty fish such as mackerel, sardines, salmon, and bluefish
Docosahexaenoic	C22:6	Fatty fish such as mackerel, sardines, salmon, and bluefish

* Carbon chain length: double bonds.

† By convention, named for the location of the first double bond from the methyl position on the carbon chain.

the landmark Seven Countries study. In that study, there were also significant associations with dietary cholesterol and trans fatty acids, but independent effects could not be shown due to high correlations between those dietary components. A recent study from eastern Finland with its excessive CHD mortality noted that cohorts from the past 20 years who had decreased their intake of high-fat liquid dairy products and spreadable fats had lower rates of CHD (Kromhout et al, 1995).

> ***Diet Counseling Tip:*** Encourage patients to read food labels carefully to determine sources of saturated fat. Watch for hidden fat in muffins and baked goods. Ask every patient to give examples of foods high in saturated fats.

■ Trans Fatty Acids

Trans fatty acids (TFAs) also raise cholesterol. The most common TFA is the trans isomer of oleic acid, elaidic acid. Its structure is such that it is straighter than the usual cis isomer and thus can be more tightly packed and hence solid at room temperature. Common sources include:

- Stick margarines and shortenings
- Dairy items such as milk, butter, and cheese
- Commercially processed baked goods.

Fast-food restaurants partially hydrogenate vegetable oils, which increases their content of TFA. Feeding studies show that TFA in usual amounts in the American diet do not raise cholesterol as much as saturated fats. Nonetheless, in increased amounts, they raise LDL-c and lower HDL-c. Many physicians and patients have been confused by the controversy.

> ***Diet Counseling Tip:*** In terms of the potential of raising blood cholesterol, butter, which contains both dietary cholesterol and saturated fat, is the most harmful, followed closely by stick margarine with its higher content of TFAs. On the other hand, small amounts of soft tub or liquid margarine should impact the lipid profile in minimal fashion.

■ Polyunsaturated Fats

Polyunsaturated fats (n-6) are essential fatty acids; thus a diet completely devoid of these fats is not healthy. These fats are represented by the seed oils and are components of soft margarines. They lower blood cholesterol and LDL-c in relation to their intake in the diet. No population has a long-term experience with high amounts of polyunsaturated oils, so their usage is limited to less than 10% of energy. Potential side effects from consuming excessive amounts of polyunsaturated oils include:

- Obesity (fats have 9 kcal/g)
- Gallstones
- Lower HDL-c levels.

Although excess polyunsaturated fat intake is associated with tumor growth in animals, this has not been seen in the many clinical studies of humans.

■ Monounsaturated Fats

Monounsaturated fatty acids (MUFAs) have become the fatty acids of choice for replacing saturated fatty acids in the diet. The chief monounsaturated fatty acid is oleic acid. Common sources include:

- Olive oil
- Canola (rapeseed) oil
- Peanut oil
- Avocados
- Almonds.

In addition to lowering blood cholesterol and LDL-c by replacing saturated fats in the diet, MUFAs appear to make LDL less susceptible to oxidation. This tendency may be particularly advantageous in those with combined hyperlipidemia and diabetes where small dense LDL are more likely to be present.

> ***Diet Counseling Tip:*** Advocating a Mediterranean-type diet is reasonable for many patients. In this diet, the source of fats is almost completely olive oil. Fatty sources of fish, poultry, and red meat are eaten less frequently. Fruits and vegetables and whole grain products are abundant. This diet may be a useful approach for the diabetic with low HDL-c and high triglycerides on a high carbohydrate diet. One potential difficulty is for the patient who overdoes the use of olive oil and gains weight on this diet!

■ Low-Fat, High-Carbohydrate Diets

Low-fat diets with high intakes of complex carbohydrate are typical of an Asian-style diet. These permit levels of fat intake as low as 10% to 15% of total energy. When these diets are introduced gradually, the marked carbohydrate-induced hypertriglyceridemia that occurs with an abrupt shift in diet may not be seen (Ullmann et al, 1991). One of the problems with these diets is the decline in HDL-c that occurs. Although the implications of HDL-c lowered in this way may be different than in those patients who have low levels of HDL-c for other reasons, it is of concern. Two strategies of clinical importance are weight loss and exercise. Tufts investigators compared a low-fat, high complex carbohydrate diet with body weight kept constant to one in which there was decreased caloric intake permitting weight loss. Only in the circumstance in which there was weight loss was the overall change in LDL-c/HDL-c favorable. Regular aerobic exercise is a useful strategy to raise HDL-c and would seem to be particularly important in this setting. When aerobic exercise was added to a prudent weight-reducing diet, HDL-c levels in men increased by approximately 13% and the expected decline in HDL-c with diet was prevented.

■ Dietary Fiber

Practically speaking, there are two kinds of fiber or nondigestible carbohydrate. The first, which is insoluble, aids in bowel function. An example is wheat bran. The second is soluble fiber, which has an additional cholesterol-lowering effect. Examples include:

- Dried beans
- Grains
- Certain fruits
- Vegetables.

Psyllium is a source of soluble fiber used as a dietary supplement or in dry cereals.

> ***Diet Counseling Tip:*** Oat products have the most soluble fiber of any grain. A simple way to improve the diet is to suggest that oatmeal, oat bran, or any of the single or multigrain dry cereals be substituted for higher-fat breakfast items. The decrease in saturated fat and increase in soluble fiber from this meal alone can produce additional lowering of LDL-c that may make increases in drug therapy unnecessary in order to reach LDL-c goals.

■ Vegetable Protein

A meta-analysis of clinical studies involving soy protein showed a small benefit in hypercholesterolemic subjects. In a recent, double-blind randomized trial in hypercholesterolemic subjects, isolated soy protein with 62 mg of isoflavones lowered

LDL-c by 4%. Ethanol-extracted isolated soy protein does not significantly lower LDL-c. Commercial products vary greatly in isoflavone content. Certainly soy is not new to the diet. Some Asian populations have consumed soy protein daily for centuries. Further work is needed, however, to establish the benefits and negative aspects of large amounts of soy in CHD risk reduction.

> ***Diet Counseling Tip:*** Lunch is a difficult meal for people on a lower-fat diet. The use of vegetable based "burgers" utilizing soy protein offers a way to get fast, convenient sandwich food which is consistent with a low saturated fat diet.

Other Dietary Factors

8

■ Coffee

Boiled coffee can raise total and LDL-c, whereas usual amounts of filtered caffeinated coffee do not. There is disagreement as to whether decaffeinated coffee raises LDL-c. Moderate amounts of filtered coffee or decaffeinated coffee do not appear to increase the risk of CHD.

> ***Diet Counseling Tip:*** When patients ask about coffee, it is important to counsel that it is what the patient has with the coffee that is often more detrimental to their coronary risk profiles. Patients need to be told to avoid nondairy coffee creamers with cholesterol-raising ingredients, such as coconut oil or palm kernel oil, to use skim milk instead of whole milk or cream, and to avoid large amounts of sugar. It has been pointed out that cappuccino with skim milk is even more frothy than that with whole milk!

■ Garlic

Although a meta-analysis suggested a small benefit of garlic on cholesterol levels, recent double-blind trials have not shown significant benefit of garlic over placebo on HDL subclasses, Lp(a), apolipoprotein B, postprandial triglycerides, or LDL subclass distribution.

> ***Diet Counseling Tip:*** Patients must consider the price of garlic supplements in the same way they consider cost of drug therapy. If significant lowering of LDL-c is required, use of a low dose of an HMG-CoA reductase inhibitor and diet has the advantage of proven efficacy and safety in high-risk populations and would be preferred over garlic supplements for more modest degrees of cholesterol lowering.

■ Excess Caloric Intake

Studies have shown that those on a high-calorie, high-fat diet increase total cholesterol, LDL-c, and HDL-c. As noted before, when calories are reduced to promote weight loss, there is often a lowering of LDL-c and HDL-c. Triglycerides are particularly sensitive to weight gain and loss and often will show the most dramatic changes. One advantage of very low fat diets, as espoused by Ornish and Pritikin, is that weight loss often occurs. For example, in the Lifestyle Heart Trial, patients assigned to the less than 10% fat, vegetarian-style diet lost 10 kg in weight over the first year. Thus the effects of weight loss must be considered as an important factor in the striking lipid changes produced with these diets.

> ***Diet Counseling Tip:*** A diet that is fat free and low calorie is not necessarily a healthful diet. Dietary strategy may be used to advantage in

those people with combined hyperlipidemia who are often middle-aged, overweight, and sedentary, and who have CHD or are greatly at risk. Lower-risk segments of the population, such as adolescents, young women, and the elderly, may pursue these diets to their disadvantage if they cause deficiencies of calcium, iron, and protein.

■ Alcohol

In cross-population studies, an increase in a patient's consumption of alcohol has been associated with reductions in CHD. Alcohol has several beneficial actions, including raising HDL-c, and potentially beneficial effects on thrombotic susceptibility involving actions on plasminogen activator, fibrinogen, and platelets. The advantages of red over white wine are not confirmed by either epidemiologic studies or studies looking at LDL susceptibility to oxidation. Thus a large part of the beneficial effect is alcohol's ability to raise HDL-c. On the other hand, there are significant negatives to recommending alcohol consumption to the public to reduce CHD risk. Increased amounts of alcohol intake over time can lead to an increase in triglycerides. In patients with familial lipid disorders where triglyceride abnormalities are prominent, alcohol consumption can greatly increase triglyceride values. In fact, excess alcohol intake is a common cause of secondary hyperlipidemias. In addition to concerns regarding alcohol and substance abuse, chronic, heavy drinking can exacerbate hypertension and lead to increased rates of:

- Stroke
- Cardiac dysrhythmia
- Left ventricular hypertrophy
- Cardiomyopathy
- Liver disease

- Accidents and suicide. (Those who consume five or more drinks per occasion are nearly twice as likely to die from injuries than persons who drink less.)

Thus there can be no public health recommendation for alcohol usage to reduce CHD risk.

> ***Diet Counseling Tip:*** Small amounts of alcohol with meals on the order of 1 or 2 units or drinks daily for men and one-half to one drink daily for women are associated with lower rates of CHD and do not need to be proscribed in adults who have shown they can control their intake. A drink is defined as $3^{1}/_{2}$ oz wine, 12 oz beer, or 1 oz hard liquor.

> ***Diet Counseling Tip:*** Suspect alcohol excess exacerbating a familial lipid disorder in men who present with severe hypertriglyceridemia and acute pancreatitis. After recovery, elevated lipids remain as a clue to the underlying lipid disorder.

> ***Diet Counseling Tip:*** Clues to increased intake of alcohol on a chronic basis include a high mean corpuscular volume, abnormal transaminase values, high serum iron, and the picture of a high triglyceride associated with a high HDL-c. Usually those with high triglycerides have lower HDL-c values.

■ Fish Oils and Fish Intake

Fish oils and fish intake are rich in the long chain n-3 polyunsaturated fatty acids (PUFAs) known as eicosapentaenoic acid (EPA) or docosahexaenoic acid (DHA). The amount of n-3 PUFA is determined by summing up the EPA and DHA of a product. Good sources of marine n-3 PUFA include:

- Salmon
- Mackerel
- Blue fish
- Sardines (watch out for the salt!).

Alpha linolenic acid (ALA) is a plant-based n-3 PUFA. Humans can synthesize some EPA and DHA from ALA. Good sources of ALA include:

- Flaxseed and canola oil
- Soybeans
- Tofu
- English walnuts.

Feeding studies have documented the great efficacy of high-dose fish oil in lowering severe hypertriglyceridemia (Chapter 10, *The Lipid-Lowering Drugs*). In high doses (above 3 g), fish oil can elevate LDL-c in subjects with diabetes or combined hyperlipidemia. It can also elevate blood glucose levels in diabetics. High doses may lower blood pressure slightly and prolong bleeding times. Low doses (under 3 g/d) are generally safe in that they do not produce the significant metabolic effects seen with higher dose fish oil. One study of fish added to a Step-II diet noted immunologic changes of uncertain significance.

Fish intake in epidemiologic studies appears to vary inversely with CHD risk, although this is not invariably the case. One large prospective study with careful dietary assessment noted a reduction in nonfatal myocardial infarction (MI), while others have suggested a reduction in sudden death. There are emerging data suggesting that a diet that includes a mild increase in n-3 PUFA may reduce sudden death:

- The Diet and Reinfarction Trial (DART) noted a 29% reduction in death, although not in reinfarction, in postinfarction male survivors randomized to a high-fish diet (25% were given low-dose fish oil capsules if they could not eat fish).

- The Lyon Diet Heart Study compared in MI survivors a Mediterranean style diet low in cholesterol and saturated fat with increased monounsaturated and n-3 PUFA compared with a control diet high in cholesterol and saturated fat. The experimental group increased their n-3 PUFA primarily through a canola-based margarine. Alcohol intake was the same in the two diet groups. There was a striking reduction in CHD death in the treatment group that was maintained at the 47-month follow-up.
- A Seattle case-control study measured n-3 PUFA in sudden death victims and found that victims were characterized by lower intakes of fish. Their data suggested a small amount of fish weekly could provide significant benefits.
- The Gruppo Italiano per lo Studio della Sopravvivenza nell'Infarto miocardico (GISSI) Prevenzione trial in over 11,000 MI survivors showed that low-dose (850 mg of n-3 PUFA) fish oil capsules reduce significantly the combined primary end point of death, nonfatal MI, and nonfatal stroke as compared with placebo. Vitamin E had no significant effect.
- Electrophysiologic studies in animals and possibly in humans have shown that n-3 PUFA increases the arrhythmia threshold for ischemia-induced sudden death.

While the data do not yet show conclusively that low-dose fish oil capsules should be taken to reduce sudden death, they do support an increase in fish meals per week (ie, 2 per week) or an increase in vegetable-based n-3 PUFA in those with CHD.

> ***Diet Counseling Tip:*** Fish is a good source of dietary protein that is low in saturated fat. Patients should be counseled to avoid rich sauces or butter when they eat fish. When eating out, they need to tell the waiter in advance how they wish to have the fish prepared and should ask for sauces on the side.

■ Antioxidants

There are many sources of antioxidants in the diet, including:

- Monounsaturated fats
- Vitamin E (alpha-tocopherol)
- Vitamin C (ascorbic acid)
- Beta carotene
- Flavonoids.

Vitamin E and beta carotene are lipid soluble, whereas vitamin C is water soluble. It is not easy to obtain large amounts of vitamin E from the diet. Food sources of vitamin E include:

- Seed oils
- Nuts
- Avocados
- Whole grain and fortified cereals
- Eggs
- Green vegetables.

Food sources of beta carotene include:

- Carrots
- Broccoli.

Vitamin C is widely available in fruits and vegetables, but loss of vitamin C can occur during cooking. Flavonoids are another source of dietary antioxidants that occur naturally in foods such as:

- Vegetables

- Fruits
- Beverages such as tea and wine.

On a milligram-per-day basis, the intake of the antioxidant flavonoids still exceeds that of the antioxidants beta carotene and vitamin E. Thus flavonoids represent a potentially important source of dietary antioxidants. These antioxidants may make LDL less susceptible to oxidation.

Until recently, there were no randomized clinical trials of antioxidant therapy. Now there are three randomized, controlled clinical trials examining the effects of vitamin E supplementation for secondary prevention. The Cambridge Heart and Antioxidant Trial (CHAOS) showed that either 800 or 400 IU of vitamin E (alpha tocopherol) reduced rates of nonfatal MI as contrasted with placebo. There was no significant effect on total mortality. The GISSI Prevention Trial (see above) and the Heart Outcomes Prevention Evaluation (HOPE) study together examined the effects of vitamin E in more than 20,000 subjects post-MI. Vitamin E did not significantly improve the primary end points compared with placebo. The almost reflex prescription of vitamin E for secondary prevention needs to be reexamined critically. Likewise, three large, randomized clinical trials have shown no benefit in CHD prevention for beta carotene. Indeed, in former smokers, it may increase the risk of lung cancer.

A final note: Increased dietary intake of carotenoids, particularly lutein and zeaxanthin, may reduce the risk of age-related macular degeneration. Increased intake of dark green leafy vegetables such as spinach and collard greens is associated with a substantial reduction in age-related macular degeneration.

> ***Diet Counseling Tip:*** Ask patients to add fruits and vegetables to their diet on a daily basis as a good way of adding antioxidants; this may prove cheaper than supplements while greatly improving the quality of the diet.

■ Homocyst(e)ine, Folic Acid, B_6 and B_{12}

Homocysteine (H[e]) is a thiol-containing amino acid which requires folic acid, vitamins B_6, and B_{12} for remethylation. By convention, the free and protein-bound forms of the amino acid and derived disulfides are called H(e). Elevated levels of H(e) may occur due to an enzyme deficiency involving cystathionine β-synthase, deficiency of 5,10-methylene tetrahydrofolate reductase, or dietary deficiencies of folic acid, B_{12}, and B_6. The latter point may be important since cross-sectional analysis of H(e) levels and vitamin levels in elderly subjects in the Framingham Study showed that plasma H(e) levels exhibited a strong inverse association with plasma folate. Moreover, a 14-year follow-up of 80,082 women using a food questionnaire showed that daily intake of 400 μg of folate and 3 mg of vitamin B_6 was associated with the lowest risk of CHD. This intake can be obtained either through fortified cereals or a multivitamin (Rimm et al, 1998). The high H(e) levels have potentially deleterious effects on endothelial cells, platelets, and smooth-muscle cells, which could promote a prothrombotic, atherogenic state. Clinical studies have shown those cohorts with high levels of H(e) have increased rates of atherosclerosis involving coronary, cerebral, or peripheral arterial vessels. Among patients with hyperlipidemia and with premature familial hyperlipidemia, elevated levels of H(e) are independent predictors of atherosclerotic events.

Due to increasing public awareness, many patients are asking for H(e) measurements and treatment regimens. The advisability of measuring H(e) levels and acting on elevated values depends on data from randomized, placebo-controlled clinical trials. These data are not yet at hand but are awaited with interest. For the general population, a recommendation for measurement of H(e) is unwarranted. For those with atherosclerotic vascular disease or familial premature CHD unexplained by other risk factors or those at high risk due to multiple risk factors, the measurement of H(e) and the recommendation of a multiple vitamin supplement containing 400 μg/d of folic acid with small amounts of B_{12} and B_6 for those with elevated values may be considered. Higher doses of folic acid are required in chronic renal failure.

> ***Clinical Tip:*** Avoid folate supplementation alone in anorectic patients or the elderly with B_{12} deficiency for whom addition of folic acid could mask the signs of B_{12} deficiency.

> ***Diet Counseling Tip:*** The awaited uniform supplementation of the food supply will probably help reduced folate deficiency.

■ Vegetarian Diets

The use of a vegetarian diet is entirely consistent with a healthful, nutritional approach to reducing risk factors for CHD. Among vegetarians, levels of both LDL-c and HDL-c are low. An informative trial in India involved advising patients recovering from an acute MI to eat more fruit, vegetables, nuts, and grain products in addition to the usual post-MI fat-reduced diet. The improvements in weight and LDL-c levels were paralleled by improvements in rates of coronary events.

Diet Counseling Tip: Patients who wish to adopt a true vegetarian lifestyle need nutritional counseling to be sure that they obtain adequate sources of nutrients. For example, vegans who eat no eggs or milk products need to provide for adequate protein and vitamin B_{12}. This is particularly important if they take folic acid supplements. Fortified soy milk may be useful. Fortified cereals can be a source of vitamin D. Ovolacto-vegetarians need to choose low-fat dairy products to keep the diet low in saturated fat.

A major problem for most physicians is finding the time to provide positive, useful messages on diets. A useful exercise is to review the sources of fat in the diet with the patient; this is the basis for the Meat, Egg yolks, Dairy products, Invisible fats, Cooking and table fats, and Snacks (MEDICS) questionnaire listed in Table 8.4. With practice, the patient can readily ascertain the sources of fat in the diet from meats, eggs, dairy products, invisible fats in baked and fried foods, cooking and table fats such as spreads and salad dressings, and snacks. The patient can review this sheet while waiting to see the physician and circle those items that require discussion. This policy allows the physician to highlight specific steps the patient can take to improve his or her nutritional patterns. For example, a diet high in fast-food items suggests that the topic of "eating out" should be a priority for the patient and dietitian. For others, tips on how to make a lower-saturated-fat lifestyle more interesting in terms of food choices, preparation, and involvement of ethnic cuisine should receive more attention. When the physician is able to endorse a healthier eating style in a positive manner, patient satisfaction and compliance may improve. In addition, the physician needs to review how the patient is do-

TABLE 8.4 — MEDICS QUESTIONNAIRE TO REVIEW SOURCES OF FAT IN THE DIET

Category	Ask Regarding	Skills to Learn
Meat	Beef, pork, lamb, or veal?	Learn to order leaner cuts of meats
	Portion size?	Restrict to lean portions of 4 oz or less
	Liver or organ meats?	Avoid organ meats
	Fowl or fish meals?	Use skinless poultry; avoid fried items
	Hot dogs or sausages?	Eat more sparingly
Egg yolks	How many per week? (4 for Step I, 2 for Step II)	Use egg whites for omelettes and recipes; consider egg substitutes
Dairy products	Whole milk products?	Use skim milk or 1% milk
	Cheeses?	Use low-fat cheeses like mozzarella
	Ice cream?	Use no- or low-fat yogurts
	Cream cheese?	Use no-fat cream cheese

Invisible fats		
Baked goods	Doughnuts, coffee cakes, pies, muffins, cakes, cookies?	Read labels and use no-fat items; watch muffins, which can be high-fat; bake with low-fat recipes
Fried foods	French fries?	Substitute fruit or pretzels for snack dips or French fries
Cooking/table fats	Oils, spreads or creamy salad dressings?	Use nonstick sprays; soft margarine instead of butter; unsaturated oils sparingly
Snacks	Ice cream, cake, candy bars, cookies?	Consider pretzels if salt is not an issue
	Snack chips?	Use more fruits and vegetables as snacks
Spirits	How many drinks of alcohol per week? (A unit or drink is 1 oz of hard liquor, one glass of wine, one can of beer)	Limit alcohol to 1 to 2 drinks per day; restrict if triglycerides are high

Abbreviations: MEDICS, Meat, Egg yolks, Dairy products, Invisible fats, Cooking and table fats, and Snacks.

8

ing periodically. Those who are having more skinless fowl and leaner cuts of red meat, avoiding high-fat salad dressings and fried foods, and snacking on fruits and vegetables have clearly made a commitment to a more healthful eating style.

SUGGESTED READINGS

Anderson JW, Johnstone BM, Cook-Newell ME. Meta-analysis of the effects of soy protein intake on serum lipids. *N Engl J Med.* 1995;333:276-282.

Criqui MH, Ringel BL. Does diet or alcohol explain the French paradox? *Lancet.* 1994;344:1719-1723.

Crouse JR 3rd, Morgan T, Terry JG, Ellis J, Vitolins M, Burke GL. A randomized trial comparing the effects of casein with that of soy protein containing varying amounts of isoflavones on plasma concentrations of lipids and lipoproteins. *Arch Intern Med.* 1999;159: 2070-2076.

de Lorgeril M, Renaud S, Mamelle N, et al. Mediterranean alpha-linolenic acid rich diet in secondary prevention of coronary heart disease [published erratum appears in *Lancet.* 1995;345:738]. *Lancet.* 1994;343:1454-1459.

Denke MA. Cholesterol-lowering diets. A review of the evidence. *Arch Intern Med.* 1995;155:17-26.

Denke MA. Review of human studies evaluating individual dietary responsiveness in patients with hypercholesterolemia. *Am J Clin Nutr.* 1995;62:471S-477S.

Denke MA, Grundy SM. Individual responses to a cholesterol-lowering diet in 50 men with moderate hypercholesterolemia. *Arch Intern Med.* 1994;154:317-325.

Fielding CJ, Havel RJ, Todd KM, et al. Effects of dietary cholesterol and fat saturation on plasma lipoproteins in an ethnically diverse population of healthy young men. *J Clin Invest.* 1995;95:611-618.

Fried RE, Levine DM, Kwiterovich PO, et al. The effect of filtered-coffee consumption on plasma lipid levels. Results of a randomized clinical trial. *JAMA.* 1992; 267:811-815.

Gardner CD, Kraemer HC. Monounsaturated versus polyunsaturated dietary fat and serum lipids. A meta-analysis. *Arterioscler Thromb Vasc Biol.* 1995;15:1917-1927.

Ginsberg HN, Karmally W, Siddiqui M, et al. A dose-response study of the effects of dietary cholesterol on fasting and postprandial lipid and lipoprotein metabolism in healthy young men. *Arterioscler Thromb.* 1994;14:576-586.

Gruppo Italiano per lo Studio della Sopravvivenza nell'Infarto micoardico. Dietary supplementation with n-3 polyunsaturated fatty acids and vitamin E after myocardial infarction: results of the GISSI-Prevenzione trial. *Lancet.* 1999:354:447-455.

Yusuf S, Dagenais G, Pogue J, Bosch J, Sleight P. Vitamin E supplementation and cardiovascular events in high-risk patients. Heart Outcomes Prevention Evaluation Study Investigators. *N Engl J Med.* 2000;342:154-160.

Hennekens CH, Buring JE, Manson JE, et al. Lack of effect of long-term supplementation with beta carotene on the incidence of malignant neoplasms and cardiovascular disease. *N Engl J Med.* 1996;334:1145-1149.

Hopkins PN, Wu LL, Wu J, et al. Higher plasma homocyst(e)ine and increased susceptibility to adverse effects of low folate in early familial coronary artery disease. *Arterioscler Thromb Vasc Biol.* 1995;15:1314-1320.

Jackson R, Beaglehole R. Alcohol consumption guidelines: relative safety vs absolute risks and benefits. *Lancet.* 1995;346:716.

Jousilahti P, Vartiainen E, Tuomilehto J, Puska P. Twenty-year dynamics of serum cholesterol levels in the middle-aged population of Eastern Finland. *Ann Intern Med.* 1996;125:713-722.

Krauss RM, Dreon DM. Low-density-lipoprotein subclasses and response to a low-fat diet in healthy men. *Am J Clin Nutr.* 1995;62:478S-487S.

Kromhout D, Menotti A, Bloemberg B, et al. Dietary saturated and trans fatty acids and cholesterol and 25-year mortality from coronary heart disease: the Seven Countries Study. *Prev Med.* 1995;24:308-315.

Lichtenstein AH, Ausman LM, Carrasco W, Jenner JL, Ordovas JM, Schaefer EJ. Short-term consumption of a low-fat diet beneficially affects plasma lipid concentrations only when accompanied by weight loss. Hypercholesterolemia, low-fat diet, and plasma lipids. *Arterioscler Thromb.* 1994;14:1751-1760.

Omenn GS, Goodman GE, Thornquist MD, et al. Effects of a combination of beta carotene and vitamin A on lung cancer and cardiovascular disease. *N Engl J Med.* 1996;334:1150-1155.

Position paper on trans fatty acids. ASCN/AIN Task Force on Trans Fatty Acids. American Society for Clinical Nutrition and American Institute of Nutrition. *Am J Clin Nutr.* 1996;63:663-670.

Rimm EB, Willett WC, Hu FB, et al. Folate and vitamin B_6 from diet and supplements in relation to risk of coronary heart disease among women. *JAMA.* 1998;279:359-364.

Ripsin CM, Keenan JM, Jacobs DR Jr, et al. Oat products and lipid lowering. A meta-analysis [published erratum appears in *JAMA.* 1992;268:3074]. *JAMA.* 1992;267;3317-3325.

Stephens NG, Parsons A, Schofield PM, Kelly F, Cheeseman K, Mitchinson MJ. Randomized controlled trial of vitamin E in patients with coronary disease: Cambridge Heart Antioxidant Study (CHAOS). *Lancet.* 1996;347:781-786.

Stone NJ. Fish consumption, fish oil, lipids, and coronary heart diseases. *Circulation.* 1996;94:2337-2340.

Superko HR, Krauss RM. Garlic powder, effect on plasma lipids, postprandial lipemia, low-density lipoprotein particle size, high-density lipoprotein subclass distribution and lipoprotein(a). *J Am Coll Cardiol.* 2000:35:321-326.

Ullmann D, Connor WE, Hatcher LF, Connor SL, Flavell DP. Will a high-carbohydrate, low-fat diet lower plasma lipids and lipoproteins without producing hypertriglyceridemia? *Arterioscler Thromb.* 1991;11:1059-1067.

9 Exercise and Lipids

Exercise training with physical fitness is clearly associated with a decrease in the risk of cardiac events and also in total mortality (Blair et al, 1989, 1995). A recent review notes the increasing evidence that regular activity of a moderate intensity (17 to 29 kJ/min, 4 to 7 kcal/min), performed by men and women over a broad age range, reduces cardiovascular mortality rates (Roy et al, 1999). This chapter will focus, however, on the role of exercise in improving the lipid profile, considering carefully the mechanisms by which this occurs.

It is important to differentiate between the effect of an acute episode on lipids from that of a prolonged repetitive series of exercises over several days/weeks/months (exercise training) (Table 9.1). An acute episode of exercise may transiently change the level of lipid and lipoproteins in the blood. Prolonged exercise training may have a longer lasting effect and be particularly useful in reducing overall coronary risk.

There is wide agreement that acute exercise lowers triglyceride levels (Oscai et al, 1972). A recent meta-analysis examined 31 randomized, controlled trials of aerobic and resistance exercise training with normolipidemic and hyperlipidemic participants that were conducted over a minimum of 4 weeks and looked at lipid and lipoprotein fractions. Aerobic exercise training in previously sedentary adults resulted in small but statistically significant decreases in total cholesterol, low-density lipoprotein cholesterol (LDL-c), and triglyceride, with an increase in high-density lipoprotein cholesterol (HDL-c) (Halbert et al, 1999). Long-distance runners and other endurance athletes generally have elevated levels of HDL-c, es-

TABLE 9.1 — EFFECTS OF EXERCISE: ACUTE EXERCISE AND CHRONIC TRAINING		
	Effect of Exercise	
	Acute Exercise	**Chronic Training**
Triglycerides	Decrease	Decrease
Cholesterol	No major change	Little to no change
LDL-c	Decrease	Minimal to no change
HDL-c	Increase	Increase
Abbreviations: LDL-c, low-density lipoprotein cholesterol; HDL, high-density lipoprotein cholesterol.		
Durstine JL, Haskell WL. *Exerc Sport Sci Rev*. 1994;22:477-521; Oscai LB, et al. *Am J Cardiol*. 1972;30:775-780.		

pecially the HDL_2 fraction. Previously sedentary men elevate HDL-c modestly after 8 to 11 months of exercise training primarily by prolonging HDL survival (Thompson, 1988). In addition, there is evidence that exercise training is associated with a change in particles away from small, dense LDL-c (considered to be atherogenic) to larger and lighter LDL (Beard, 1996) (Table 9.2).

Exercise may benefit those who undertake the National Cholesterol Education Program (NCEP) Step I or II diets. A recent meta-analysis showed that exercise resulted in greater decreases in total cholesterol, LDL-c, and triglyceride and prevented the decrease in HDL-c associated with low-fat diets (Yu-Poth et al, 1999). In fact, in one study, of those with low HDL-c and high LDL-c, the NCEP Step II diet failed to lower LDL-c levels in men or women who did not engage in aerobic exercise (Stefanick et al, 1998). Finally, initial trial data suggested that exercise helped ameliorate the reduction in HDL-c in overweight and obese women that occurred when they were on weight reduction diets (Wood et al, 1991). Nonetheless, the

TABLE 9.2 — EFFECTS OF DIET AND EXERCISE ON LIPIDS OVER 6 WEEKS			
	Intervention (mg/dL)		Change (%)
	Pre-	Post-	
Cholesterol	233 ± 5.4	186 ± 4	– 20
LDL-c	135 ± 4	109 ± 3	– 20
HDL-c	54 ± 2	47 ± 1	– 17
Triglycerides	221 ± 13	159 ± 9	– 26
LDL particle diameter	24.2 ± 0.2	25.1 ± 14	—

Abbreviations: LDL-c, low-density lipoprotein cholesterol; HDL-c, high-density lipoprotein cholesterol.

Beard CM, et al. *Arterioscler Thromb Vasc Biol.* 1996;16:201-207.

data set is not large and some believe too limited to determine whether physical activity can raise low HDL-c or lower high triglyceride or LDL-c levels in overweight and obese individuals (Stefanick, 1999).

The magnitude of changes in lipid levels that can be expected from exercise is dependent on several factors, including:

- Degree of abnormality before intervention
- Type of intervention
- Associated interventions
- Underlying genetic milieu.

In general, the more abnormal the lipid levels before the intervention is begun, the greater the potential beneficial change. Therefore, if the lipids are close to normal, a small change may be all that will be attained. However, this may be all that is needed. The majority of the studies have used either bicycle exercise or running to achieve the exercise state. Some studies have suggested that moderate-intensity weight training may induce similar changes. This is in contrast to the observation that body builders tend to have less normalization of lipids than weight-matched controls.

The changes in lipids and lipoprotein levels with exercise appear to be dose dependent. Every patient, however, cannot be a long-distance runner. While all patients can benefit from moderate-intensity exercise in terms of improved fitness, in older patients the change in HDL-c seen may be small and require a longer time frame (2 years) than reported previously for younger populations (King et al, 1995). A key variable in achieving a higher HDL-c is the frequency of participation. A supervised, home-based exercise regimen can be a safe and attractive alternative for achieving the sustained participation required to give best results in raising HDL-c. In a recent comparison of lifestyle physical activity intervention vs a structured exercise program, in sedentary adults there were comparable improvements in physical activity, cardiorespiratory fitness, and blood pressure but not in HDL-c. After 24 months, however, the ratio of total cholesterol to HDL-c increased significantly in both groups ($P = 0.03$ for lifestyle and $P < 0.002$ for structured exercise) (Dunn et al, 1999).

Thus prescribing exercise for 5 days per week for between 30 to 45 minutes will likely provide health benefits for your patients, albeit small changes initially in lipid levels. This is an energy expenditure of approximately 2 kcal/kg/d or about 150 kcal/d for a 75-kg person which translates to approximately 10 miles of brisk walking for 1 week (Dunn et al, 1999).

The addition of specific dietary interventions to exercise may also influence the type of change in lipid or lipoprotein levels seen with exercise. For example, if severe fat restriction is added to exercise, there will be a greater change in cholesterol and triglyceride levels than if diet is not added to exercise. However, the HDL-c levels will probably go down instead of up as they will if there is no change in diet. Another study showed that on a "normal" diet, a walking program led to an increase in HDL-c and a fall in very-low-density lipoprotein cholesterol (VLDL-c). On a high carbohydrate diet, the same walking program was associated with a fall in HDL-c levels and an increase in VLDL-c levels, along with a fall in LDL-c levels. On a high-fat diet, there was an increase in HDL-c and a fall in VLDL-c with a fall in VLDL/triglyceride levels.

There is evidence that the types of apoprotein molecules in the lipid particles themselves have a relationship to the effect of exercise on lipid and lipoprotein levels. One of the first apoproteins to be implicated in governing the response of lipoproteins to exercise is the apo E apoprotein. For example, in young subjects with apo E phenotype E-IV/IV, physical activity appears to have minimal if any effect on total cholesterol, LDL-c, or HDL-c levels. On the other hand, people with the phenotype E-III/II had the most marked reduction in triglyceride and LDL-c levels and an increase in HDL-c levels with exercise training. Subjects with other phenotypes had responses between those with type E-IV/IV and E-III/II.

A single episode of exercise increases lipoprotein lipase (LPL) activity and chronic training appears to be associated with increases in post-heparin LPL activity. This can be accompanied by chylomicron and VLDL hydrolysis. The remnants from this activity may lead to increased conversion of HDL_3 to HDL_2. Exercise may also lead to reductions in hepatic lipase activity that would be associated with lower rates of HDL_2 breakdown. In patients with isolated low HDL-c, exercise may be useful in improving HDL-c (Table 9.3). Exercise training is associated to a reduction in cholesterol ester transfer protein which may also improve lipid abnormalities (Table 9.4).

TABLE 9.3 — EFFECTS OF EXERCISE IN SUBJECTS WITH LOW HDL-C LEVELS		
	Intervention (mg/dL)	
	Preexercise	**Postexercise**
Cholesterol	195 ± 41	192 ± 35
Triglycerides	211 ± 121	182 ± 138
HDL-c	29 ± 37	33 ± 6
LDL-c	124 ± 37	123 ± 29

Abbreviations: HDL-c, high-density lipoprotein cholesterol; LDL-c, low-density lipoprotein cholesterol.

Lavie CJ, Milani RV. *Am J Cardiol.* 1996;78:1286-1289.

TABLE 9.4 — EFFECTS OF EXERCISE TRAINING ON LIPIDS AND CHOLESTERYL ESTER TRANSFER PROTEIN		
Variable	**Control**	**Effect of Training Protocol**
VO_2 Max (mL/kg/min)	—	+ 5.3 ± 3.5
Weight (kg)	—	– 2.5 ± 3.5
TG (mg/dL)	—	– 25.7 ± 36.3
HDL-c	—	+ 2.6 ± 6.2
CETP (men)	2.47 ± 0.66	2.12 ± 0.43
CETP (women)	2.72 ± 1.01	2.36 ± 0.76

Abbreviations: VO_2, volume of oxygen consumption per unit of time; TG, triglyceride; HDL-c, high-density lipoprotein cholesterol; CETP, cholesteryl ester transfer protein.

Seip RL, et al. *Arterioscler Thromb.* 1993;13:1359-1367.

Exercise can be prescribed by:

- Frequency
- Intensity
- Time (duration).

An ideal exercise program should consist of a minimum of three and preferably five sessions a week, each lasting from 30 to 40 minutes and of an intensity designed to produce mild dyspnea and/or diaphoresis.

SUGGESTED READINGS

Beard CM, Barnard RJ, Robbins DC, Ordovas JM, Schaefer EJ. Effects of diet and exercise on qualitative and quantitative measures of LDL and its susceptibility to oxidation. *Arterioscler Thromb Vasc Biol*. 1996;16:201-207.

Blair SN, Kohn HW 3rd, Paffenbarger RS Jr, Clark DG, Cooper KH, Gibbons LW. Physical fitness and all-cause mortality. A prospective study of healthy men and women. *JAMA*. 1989;262:2395-2401.

Blair SN, Kohl HW 3rd, Barlow CE, Paffenbarger RS Jr, Gibbons LW, Macera CA. Changes in physical fitness and all-cause mortality. A prospective study of healthy and unhealthy men. *JAMA*. 1995;273:1093-1098.

Duncan JJ, Gordon NF, Scott CB. Women walking for health and fitness. How much is enough? *JAMA*. 1991;266:3295-3299.

Dunn AL, Marcus BH, Kampert JB, Garcia ME, Kohl HW 3rd, Blair SN. Comparison of lifestyle and structured interventions to increase physical activity and cardiorespiratory fitness: a randomized trial. *JAMA*. 1999;281:327-334.

Durstine JL, Haskell WL. Effects of exercise training on plasma lipids and lipoproteins. *Exerc Sport Sci Rev*. 1994;22:477-521.

Fonong T, Toth MJ, Ades PA, Katzel LI, Calles-Escandon J, Poehlman ET. Relationship between physical activity and HDL-cholesterol in healthy older men and women: a cross-sectional and exercise intervention study. *Atherosclerosis*. 1996;127:177-183.

Fried LP, Kronmal RA, Newman AB, et al. Risk factors for 5-year mortality in older adults: the Cardiovascular Health Study. *JAMA*. 1998;279:585-592.

Ginsberg GS, Agil A, O'Toole M, Rimm E, Douglas PS, Rifai N. Effects of a single bout of ultraendurance exercise on lipid levels and susceptibility of lipids to peroxidation in triathletes. *JAMA*. 1996;276:221-225.

Griffin BA, Skinner ER, Maughan RJ. The acute effect of prolonged walking and dietary changes on plasma lipoprotein concentrations and high-density lipoprotein subfractions. *Metabolism*. 1988;37:535-541.

Halbert JA, Silagy CA, Finucane P, Withers RT, Hamdorf PA. Exercise training and blood lipids in hyperlipidemic and normolipidemic adults: a meta-analysis of randomized, controlled trials. *Eur J Clin Nutr*. 1999;53:514-522.

King AC, Haskell WL, Young DR, Oka RK, Stefanick ML. Long-term effects of varying intensities and formats of physical activity on participation rates, fitness, and lipoproteins in men and women aged 50 to 65 years. *Circulation*. 1995;91:2596-2604.

Lavie CJ, Milani RV. Effects of cardiac rehabilitation, exercise training, and weight reduction on exercise capacity, coronary risk factors, behavioral characteristics, and quality of life in obese coronary patients. *Am J Cardiol*. 1997;79;397-401.

Lavie CJ, Milani RV. Effects of nonpharmacologic therapy with cardiac rehabilitation and exercise training in patients with low levels of high-density lipoprotein cholesterol. *Am J Cardiol*. 1996; 78:1286-1289.

Nicklas BJ, Katzel LI, Busby-Whitehead J, Goldberg AP. Increases in high-density lipoprotein cholesterol with endurance exercise training are blunted in obese compared with lean men. *Metabolism*. 1997;46:556-561.

Niebauer J, Hambrecht R, Schlierf G, et al. Five years of physical exercise and low fat diet: effects on progression of coronary artery disease. *J Cardpulm Rehabil*. 1995;15:47-64.

Oscai LB, Patterson JA, Bogard DL, Beck RJ, Rothermel BL. Normalization of serum triglycerides and lipoprotein electrophoretic patterns by exercise. *Am J Cardiol*. 1972;30:775-780.

Seip RL, Moulin P, Cocke T, et al. Exercise training decreases plasma cholesteryl ester transfer protein. *Arterioscler Thromb.* 1993;13:1359-1367.

Shephard RJ, Balady GJ. Exercise as cardiovascular therapy. *Circulation.* 1999;99:963-972.

Stefanick ML. Physical activity for preventing and treating obesity-related dyslipoproteinemias. *Med Sci Sports Exerc.* 1999;31 (suppl 11):S609-S618.

Stefanick ML, Mackey S, Sheehan M, Ellsworth N, Haskell WL, Wood PD. Effects of diet and exercise in men and postmenopausal women with low levels of HDL cholesterol and high levels of LDL cholesterol. *N Engl J Med.* 1998;339:12-20.

Taimela S, Lehtimaki T, Porkka KV, Rasanen L, Viikari JS. The effect of physical activity on serum total and low-density lipoprotein cholesterol concentrations varies with apolipoprotein E phenotype in male children and young adults: The Cardiovascular Risk in Young Finns Study. *Metabolism.* 1996;45:797-803.

Thompson PD, Cullinane EM, Sady SP, et al. Modest changes in high-density lipoprotein concentration and metabolism with prolonged exercise training. *Circulation.* 1988;78:25-34.

US Department of Health and Human Services, Centers for Disease Control and Prevention, and National Center for Chronic Disease Prevention and Health Promotion. Physical activity and health: a report of the surgeon general. Atlanta, Ga: Centers for Disease Control and Prevention; 1996.

Winslow E, Bonhannon N, Brunton SA, Mayhew HE. Lifestyle modification: weight control, exercise and smoking cessation. *Am J Med.* 1996;101(suppl 4A):25S-31S, discussion 31S-33S.

Wood PD, Stefanick ML, Williams PT, Haskell WL. The effects on plasma lipoproteins of a prudent weight-reducing diet, with or without exercise, in overweight men and women. *N Engl J Med.* 1991;325:461-466.

Yu-Poth S, Zhao G, Etherton T, Naglak M, Jonnalagadda S, Kris-Etherton PM. Effects of the National Cholesterol Education Program's Step I and Step II dietary intervention programs on cardiovascular disease risk factors: a meta-analysis. *Am J Clin Nutr.* 1999;69:632-646.

10 The Lipid-Lowering Drugs

Several medications have come into widespread use for altering the plasma concentrations of lipoproteins. Their mechanisms of action involve:

- Modification of lipoprotein synthesis
- Modification of intravascular metabolism of lipoproteins
- Alteration of clearance of lipoproteins (Table 10.1).

When dietary measures have proven to be insufficient, the use of these agents should be considered. The mechanisms of action and the major indications of these medications are shown in Table 10.2. The usual dosing regimens of the various lipid-lowering drugs are shown in Table 10.3.

It is important to recognize that these drugs act on lipoproteins, the transport vehicles for the plasma lipids. Thus drugs such as the bile acid sequestrants (cholestyramine and colestipol) or the HMG-CoA reductase inhibitors must be considered to be low-density lipoprotein (LDL)-reducing agents; they are *not* generic cholesterol reducers. In cases when elevated plasma cholesterol levels are not due to elevations of LDL (as in severe hypertriglyceridemia and when cholesterol is elevated because of elevated high-density lipoprotein [HDL] levels), these drugs are not appropriate and will not be effective.

HMG-CoA Reductase Inhibitors (The "Statins")

Six drugs in this class are currently in use:

- Atorvastatin (Lipitor)

TABLE 10.1 — CLASSIFICATION OF LIPID-LOWERING MEDICATIONS ACCORDING TO MAJOR MECHANISM OF ACTION	
Reduce lipoprotein synthesis/secretion	Nicotinic acid, fish oils (omega-3 fatty acids)
Alter intravascular metabolism	Fibric acid derivatives
Enhance low-density lipoprotein clearance (receptor-mediated)	Bile acid sequestrants, HMG-CoA reductase inhibitors

- Cerivastatin (Baycol)
- Fluvastatin (Lescol)
- Lovastatin (Mevacor)
- Pravastatin (Pravachol)
- Simvastatin (Zocor)

As a class, these drugs are the most easily tolerated and the most efficacious agents for reducing LDL-c levels. For this reason, they have become the most widely used lipid-altering medications.

Lovastatin, simvastatin, and pravastatin are fungal metabolites or derivatives thereof. Fluvastatin, atorvastatin, and cerivastatin are totally synthetic. Lovastatin and simvastatin are administered as inactive lactones; they must be hydrolyzed to the corresponding hydroxyacids in order to attain pharmacologic activity. Thus lovastatin and simvastatin can be considered to be prodrugs. Hydrolysis of the inactive lactones occurs within hepatocytes. Pravastatin, fluvastatin, atorvastatin, and cerivastatin are administered in active form.

■ Mechanism of Action and Drug Metabolism

The HMG-CoA reductase inhibitors act by inhibiting 3-hydroxy-3-methylglutaryl coenzyme A, the rate-limiting enzyme of cholesterol biosynthesis. This

TABLE 10.2 — LIPID-ALTERING MEDICATIONS: MECHANISMS OF ACTION AND MAJOR INDICATIONS

	Primary Metabolic Effect	Effects on Lipoprotein Metabolism	Primary Indications*
HMG-CoA reductase inhibitors	Competitively inhibits cholesterol synthesis	Increases LDL clearance via receptors	High LDL High VLDL
Bile acid sequestrants	Interrupts enterohepatic circulation of bile acids	Increases LDL clearance via receptors	High LDL
Nicotinic acid	Inhibits lipolysis in adipocytes	Decreases VLDL synthesis; decreases HDL clearance	High LDL High VLDL High IDL (type III) Low HDL
Fibric acid derivatives	Increases lipoprotein lipase activity	Enhances VLDL clearance; increases HDL synthesis	High VLDL High IDL (type III) Low HDL with elevated LDL High triglyceride with high LDL/HDL ratio
Fish oils	Increased intracellular degradation of apo B-100	Inhibits VLDL secretion	High triglyceride

Abbreviations: HMG-CoA, 3-hydroxy-3-methylglutaryl coenzyme A; LDL, low-density lipoprotein; VLDL, very low-density lipoprotein; HDL, high-density lipoprotein; IDL, intermediate-density lipoprotein; apo, apolipoprotein.

* Indicated for use in patients with lipid and/or triglyceride levels listed.

TABLE 10.3 — LIPID-LOWERING DRUGS: PREPARATIONS AND USUAL DOSING REGIMENS

Drug (Trade Name)	Availability	Starting Dose	Dose Range
Cholestyramine (Questran, Questran Light, LoCholest, LoCholest Light, Prevalite)	Powder for Oral Suspension: single-dose (4 g) packets or cans with dosing scoop (1 scoop = 1 dose)	4 g/d (1 scoop)	4-12 g bid
Colestipol (Colestid)	Granules: single-dose (5 g) packets or bottles with dosing scoop (1 scoop = 1 dose) Tablet: 1 g	5 g/d (1 scoop)	5-15 g bid
Gemfibrozil (Lopid)	Tablet: 600 mg	600 mg bid	600 mg bid
Fenofibrate (Tricor)	Gelatin Capsule: 67, 134, 200 mg	67 mg/d	67-200 mg/d
Nicotinic acid			
Immediate-release (Niacor, others)	Tablet: 5, 100, 250, 500, 1000 mg	50 mg tid	500-2000 mg tid
Sustained-release (Niaspan, Slo-Niacin, others)	Tablet: 250, 375, 500, 750, 1000 mg	500 mg/d	1000-2000 mg/d

Fish oil	Capsule: 1 g	3 g tid	2-6 g tid
HMG-CoA Reductase Inhibitors			
Atorvastatin (Lipitor)	Tablet: 10, 20, 40 mg	10 mg qpm	10-80 mg/d
Cerivastatin (Baycol)	Tablet: 0.2, 0.3, 0.4 mg	0.3 mg qpm	0.2-0.4 mg qpm
Fluvastatin (Lescol)	Capsule: 20, 40 mg	20-40 mg qpm	20-40 mg qpm, 40 mg bid
Lovastatin (Mevacor)	Tablet: 10, 20, 40 mg	20 mg qpm	10-40 mg qpm, 40 mg bid
Pravastatin (Pravachol)	Tablet: 10, 20, 40 mg	20 mg qpm	10-40 mg qpm
Simvastatin (Zocor)	Tablet: 5, 10, 20, 40, 80 mg	20 mg qpm	5-80 mg qpm

Abbreviation: HMG-CoA, 3-hydroxy-3-methylglutaryl coenzyme A.

causes a sequence of steps, having the end result of increasing LDL receptor activity on hepatocytes and thereby speeding the clearance of LDL from plasma.

Although increased clearance of LDL by LDL receptors is the most important effect of the statins, they also may reduce the production and enhance the hepatic clearance of very low-density lipoprotein (VLDL). This would account for the triglyceride reductions seen when these drugs are used.

Each of these drugs has two key structural features:

- One portion of the drug mimics the structure of coenzyme A and fits into the coenzyme A binding site of the HMG-CoA reductase enzyme.
- The other portion mimics the structure of the partially reduced intermediate produced as hydroxymethylglutarate and is converted to mevalonate.

A crucial consequence of inhibition of biosynthesis of cholesterol within hepatocytes is a reduction in intracellular cholesterol stores. Homeostatic mechanisms within the hepatocyte (involving a sterol response element in the promoter region of the LDL receptor gene) then increase LDL receptor activity on the cell membrane, and LDL is cleared more rapidly from the circulation, bringing its cholesterol content into the hepatocyte.

Gastrointestinal absorption of these drugs varies from 31% (lovastatin) to >90% (fluvastatin). All of the statins are targeted to the liver, their pharmacologic site of action, by first-pass hepatic extraction. The extent of hepatic extraction varies from a low of 20% to 25% (cerivastatin) to a high of at least 79% (simvastatin) (Table 10.4).

These drugs are extensively protein bound (>95%) except for pravastatin, for which protein bind-

TABLE 10.4 — METABOLIC PARAMETERS FOR HMG-COA REDUCTASE INHIBITORS

Drug	Absorption (%)	Plasma Protein Binding (%)	Hepatic Extraction (% of absorbed dose)	Renal Excretion (%)	Plasma Half-Life (h)
Atorvastatin	NA	NA	NA	<17	11-24
Cerivastatin	>80	>99	20-25	24-30	2-3
Fluvastatin	>90	>99	>68	6	0.5-0.8
Lovastatin	31	>95	>69	30	3
Pravastatin	34	43-48	46	60	3
Simvastatin	61-85	98	>79	13	NA

Abbreviations: HMG-CoA, 3-hydroxyl-3-methylglutaryl coenzyme A; NA, not applicable.

Adapted from Blum CB. *Am J Cardiol*. 1994;73:3D-11D. Data for atorvastatin from: *Physicians' Desk Reference*. 54th ed. Montvale, NJ: Medical Economics Company, Inc; 2000:2349-2352; Cilla DD Jr, et al. *J Clin Pharmacol*. 1996;36:604-609; and Cilla DD Jr, et al. *Clin Pharmacol Ther*. 1996;60:687-695. Data for cerivastatin from: Muck W, et al. *Int J Clin Pharmacol Ther*. 1997;35:255-260; and *Physicians' Desk Reference*. 54th ed. Montvale, NJ: Medical Economics Company, Inc; 2000:675-677.

ing is somewhat below 50%. It has been speculated that a high degree of protein binding may minimize the effects of these drugs on nonhepatic tissues, thereby minimizing the potential for adverse effects in nonhepatic tissues. It is not clear, however, whether these agents actually differ from one another in the frequency of such side effects.

The liver provides the major route for clearance of the statins. Significant renal excretion occurs only for pravastatin and cerivastatin. However, even for these two drugs, hepatic clearance is substantial. Renal failure does not cause elevated blood levels of pravastatin. A reduced dose of cerivastatin (0.2 mg/d) is recommended for persons with renal failure. For lovastatin, increased levels of drug activity are seen in uremic patients. After an intravenous dose of fluvastatin, only 6% appears in the urine; thus renal failure does not warrant modification of dosing with fluvastatin.

■ Efficacy—Lipid Effects

The statins are the most effective agents available for reducing the blood levels of LDL-c. Additionally, they have relatively weak HDL-increasing effects, as well as having triglyceride-lowering effects, which are most pronounced in hypertriglyceridemic patients. They have no impact on the plasma levels of Lp(a). When given as a single daily dose, the statins (except atorvastatin) produce a greater reduction in LDL if dosing is in the evening. Atorvastatin's efficacy is not influenced by the timing of its single daily dose.

Of the six drugs in this class currently in use, the effects on LDL-c, HDL-c, and triglycerides are summarized in Table 10.5. In producing this table, the largest blinded studies providing information on the response to specific doses of each drug have been used. Studies were excluded if their designs included biases such as titration of dose according to the response to treatment.

The various drugs do differ in the amount of LDL-c reduction that can be obtained with a maximum dose. The maximum approved dose of atorvastatin (80 mg/d) gives an average 58% LDL reduction in hypercholesterolemic patients. This is greater than the LDL reduction seen with the maximum approved doses of the other statins. The increased LDL reduction possible with atorvastatin appears to be related to its longer plasma half-life (Naoumova et al, 1997). This distinction among the various statins becomes important only when maximal LDL reduction is needed. Most patients who require LDL-reducing medication can achieve treatment goals at much less than a 58% reduction of LDL-c.

The statins increase HDL levels. These changes may contribute to clinical benefit seen with these drugs. Two reports of open-label studies comparing simvastatin with atorvastatin indicate that simvastatin may be the more effective in increasing HDL levels (Jones, 1998; Crouse, 1999). Crouse reported that in studies with a total of 842 subjects, simvastatin 40 mg/d caused HDL cholesterol to increase by 6.7%, while atorvastatin 20 mg/d caused change in HDL cholesterol to increase by 4%; these doses of the two drugs were similar to one another in their effects on LDL (– 42.7% and – 45.0%, respectively). A similar conclusion was drawn in comparison of simvastatin 80 mg/d with atorvastatin 40 mg/d (HDL changes of + 6.6% with simvastatin vs + 3.0% with atorvastatin, and LDL changes of – 49.2% and – 51.5%, respectively).

The statins also reduce triglyceride levels and they should be considered for treatment of moderate elevations of triglycerides if drug therapy is appropriate (Grundy, 1998). The various statins, when compared in doses giving similar cholesterol lowering, have

TABLE 10.5 — EFFICACY OF HMG-COA REDUCTASE INHIBITORS				
Drug	**Dose (mg/d)**	**% Δ in LDL-c**	**% Δ in HDL-c**	**% Δ in TG**
Atorvastatin[1]	5	– 29	+ 8	– 25
	10	– 36	+ 7	– 13
	20	– 46	+ 6	– 22
	40	– 50	+ 3	– 30
	80	– 58	+ 2	– 26
Cerivastatin[2]	0.2	– 28	+ 6	—
	0.3	– 31	+ 8	—
	0.4	– 36	+ 4	—
Fluvastatin	20[3]	– 21	+ 3	– 8
	40[3]	– 26	+ 3	– 12
	80[4]*	– 32	—	—
Lovastatin[5]	20	– 24	+ 7	– 10
	40	– 34	+ 9	– 16
	80	– 40	+ 10	– 19
Pravastatin	10[6]	– 18	+ 5	– 5
	20[7]	– 25	+ 16	– 13
	40[8]	– 28	+ 7	– 11
Simvastatin	5[9]	– 23	+ 8	– 10
	10[9]	– 28	+ 6	– 9
	20[9]	– 37	+ 6	– 12
	40[10]	– 40	+ 12	– 19
	80[10]	– 46	+ 4[†]	– 19[†]
	80[10]	– 46	+ 10[‡]	– 36[‡]

Abbreviations: HMG-CoA, 3-hydroxy-3-methyglutaryl coenzyme A; LDL-c, low-density lipoprotein cholesterol; HDL-c, high-density lipoprotein cholesterol; TG, triglyceride.

* Administered as 40 mg bid.
† TG <200 mg/dL.
‡ TG >200 mg/dL.

1. Data for 5 mg/d (n = 13) are taken from Nawrocki JW, et al. *Arterioscler Thromb Vasc Biol.* 1995;15:678-682. Data for 10 mg/d (n = 1090) are taken from: (1) Bertolini S, et al. *Atherosclerosis.* 1997;130:191-197; (2) Dart A, et al. *Am J Cardiol.* 1997;80:39-44; (3) Davidson M, et al. *Am J Cardiol.* 1997;79:1475-1481; (4) Heinonen TM, et al. *Clin Ther.* 1996;18:853-863; and (5) Nawrocki JW, et al. *Arterioscler Thromb Vasc Biol.* 1995;15:678-682. Data for 40 mg/d (n = 26) are taken from: (1) Cilla DD Jr, et al. *J Clin Pharmacol.* 1996;36:604-609; and (2) Nawrocki JW, et al. *Arterioscler Thromb Vasc Biol.* 1995;15:678-682. Data for 80 mg/d (n = 11) are taken from Nawrocki JW, et al. *Arterioscler Thromb Vasc Biol.* 1995;15:678-682. Jones P, et al. *Am J Cardiol.* 1998;81:582-587 contributes data for 10, 20, 40, and 80 mg/d.
2. Data are taken from double-blind studies (n = 641); Stein E. *Am J Cardiol.* 1998;82:40J-46J.
3. Data for 20 mg/d (n = 1066) and 40 mg/d (n = 633) are taken from a summary of blinded, placebo-controlled trials: Peters TK, et al. *Drugs.* 1994;47(suppl 2):64-72.
4. Data are taken from the largest single controlled study using this dose (n = 266): *Physicians' Desk Reference.* 54th ed. Montvale, NJ: Medical Economics Company, Inc; 2000:2021-2024.
5. Data are taken from the EXCEL Study (n = 8245): Bradford RH, et al. *Arch Intern Med.* 1991;151:43-49.
6. Data are taken from a double-blind trial (n = 138): Steinhage-Thiessen E. *Cardiology.* 1994;85:244-254.
7. Data are taken from a double-blind trial (n = 303): The Lovastatin Pravastatin Study Group. *Am J Cardiol.* 1993;71: 810-815.
8. Data are taken as mean from three clinical trials with sample size of at least 500 each (CARE, WOSCOPS, and REGRESS).
9. Data are taken from: (1) Farmer JA, et al. *Clin Ther.* 1992; 14:708-717; (2) Douste-Blazy P, et al. *Drug Invest.* 1993;6:353-361; (3) Steinhagen-Thiessen E. *Cardiology.* 1994;85:244-254; (4) Lambrecht LJ, Malini PL. *Acta Cardiol.* 1993;48:541-554.
10. Data are taken from Stein EA, et al. *Am J Cardiol.* 1998;82: 311-316.

similar effects on triglycerides. The magnitude of triglyceride reduction is dependent on baseline triglyceride levels. Thus, with a baseline triglyceride level of <150 mg/dL, triglyceride reduction is minimal. When baseline triglyceride levels are 150 to 250 mg/dL, the ratio of the percent change in triglyceride/percent change in LDL cholesterol is 0.5; when baseline triglyceride levels are >250 mg/dL, the ratio is 1.2 (Stein, 1998).

In selecting a statin, it should be recognized that it is not always necessary to achieve the maximum LDL-c reduction that these drugs can deliver. The desired impact of treatment depends on a patient's baseline LDL-c level and the LDL-c level defined as the goal of treatment. For patients with extreme elevations of LDL-c, a high dose of atorvastatin may be needed. For the larger number of patients with less severe elevations of LDL-c, other drugs and lower doses will often suffice.

■ Efficacy—Clinical and Atherosclerosis End Points

Clinical trials (reviewed in detail in Chapter 7, *Clinical Trials*) have demonstrated reduced progression of atherosclerosis or reduced coronary heart disease rates for each of the statins except for cerivastatin. Therefore, for the present, we should reserve cerivastatin for such circumstances when the other statins (atorvastatin, fluvastatin, lovastatin, pravastatin, or simvastatin) are unacceptable for reasons of efficacy or adverse effects.

The statins have reduced the incidence of clinical cardiovascular events in patients with preexisting coronary heart disease and in those without preexisting CHD. Benefits have also been demonstrated in angiographic assessment of coronary atherosclerosis. Furthermore, a reduced frequency of stroke has been demonstrated.

Similar benefits are seen when LDL-c is reduced by other drugs, by therapies such as apheresis or partial ilial bypass, or by lifestyle changes. Therefore, the benefits shown in the individual clinical trials should be viewed as being due primarily to reduction of LDL-c. These benefits (reduced coronary atherosclerosis and reduced coronary heart disease [CHD] events) should be applicable to every drug in the statin class.

However, there is a theoretic basis for the concept that the statins may provide benefit by means other than reduction of LDL. For example, by reducing the synthesis of nonsterol products of mevalonate (eg, the regulatory compound geronylgeronyl pyrophosphate), these drugs have been shown to increase the synthesis of endothelial nitric oxide synthase and to improve arterial vasodilatory function. It has been speculated that this may be responsible for stroke prevention by the statins (Laufs and Liao, 1998). Additionally, simvastatin has been shown to inhibit vascular smooth muscle cell proliferation by a similar mechanism; we may speculate that this also may help to limit atherogenesis (Laufs et al, 1999).

There has been an interesting report on the use of pravastatin in heart transplant recipients (Kobashigawa et al, 1995). Here, 97 transplant recipients were randomly assigned to receive either pravastatin or a placebo. Those receiving pravastatin had a markedly reduced incidence of severe episodes of transplant rejection (episodes with hemodynamic compromise), reduced development of transplant vasculopathy, improved survival during the first year after transplantation, and reduced natural killer–cell cytotoxicity. The finding of reduced severe rejection is particularly intriguing. Several mechanisms have been postulated, including:

- Direct immunosuppressive effects of the statins
- Influence of reduced LDL levels on cyclosporine pharmacokinetics (cyclosporine is transported in LDL) (Akhlaghi et al, 1997)
- Reduced atherosclerosis via the same mechanisms that apply to nontransplant patients treated with statins.

In another study of cardiac transplantation, simvastatin was shown to reduce the incidence of accelerated graft vessel disease and to improve survival; its use was also associated with a trend toward less frequent episodes of acute graft rejection (Wenke et al, 1997).

Recent reports indicate that statins may also have beneficial effects on bone, leading to a substantial reduction in the risk of fractures (odds ratio 0.55 in report of Meier et al, 2000; 71% in report of Wang et al, 2000). It is hypothesized that this is caused by statin-mediated reduction of synthesis of geranylgeranyl pyrophosphate, a regulatory compound that is a nonsterol product of mevalonate (Cummings and Douglas, 2000).

■ Adverse Effects

The most important adverse effects of the statins are as follows:

- Hepatotoxicity
- Myopathy
- Teratogenicity (possibly) (Table 10.6).

TABLE 10.6 — ADVERSE EFFECTS OF HMG-COA REDUCTASE INHIBITORS
Major Side Effects • Hepatotoxicity • Myopathy • Teratogenicity
Minor Side Effects • Dyspepsia • Generalized eczematous rash
Abbreviation: HMG-CoA, 3-hydroxy-3-methylglutaryl coenzyme A.

Hepatotoxicity is manifested by elevation of transaminases. It is dose-related, usually unassociated with symptoms, and slowly reversible on discontinuation of the drug. The frequency of this side effect is about 1%. Hepatotoxicity appears to be directly linked to the mechanism of action of the drug, inhibition of 3-hydroxy-3-methylglutaryl coenzyme A reductase. Mild elevations of transaminases do not warrant discontinuation of treatment. However, transaminase elevations persistently in excess of 3 times the upper limit of normal warrant discontinuation of the drug. After resolution of the transaminase elevation, resumption at a lower dose can be considered. Monitoring of transaminases is recommended for patients treated with the statins. This is particularly needed at 6 to 12 weeks after initiation of therapy or after an increase in dose. Hepatotoxicity is more likely to occur in persons who are concurrently taking other hepatotoxic drugs or who consume alcohol on a regular basis.

Myopathy, leading to extreme weakness, myalgia, and marked elevation of creatine kinase, is a rare adverse effect of the statins. Rhabdomyolysis with renal failure has been reported rarely. In the Expanded Clinical Evaluation of Lovastatin (EXCEL) Study, the frequency of myopathy was related to the dose of lovastatin, being maximal (0.24%) with a dose of 40 mg bid. There is no convincing evidence to indicate that the frequency of myopathy differs among these agents. Myopathy has been reported to be more frequent when lovastatin is used concomitantly with cyclosporin A (30% reported frequency of myopathy), gemfibrozil (5%), nicotinic acid (3%), or erythromycin (3%). However, most experts believe that these estimates markedly overstate the actual risk when the dose of statins is limited (eg, simvastatin up to 20 mg/d). Kobashigawa and associates reported no episode of myopathy in 47 heart transplant recipients who were treated with both pravastatin and cyclosporin A. Similarly, Wenke and co-workers noted no episode of myopathy in 35 heart transplant patients who were treated with both simvastatin and cyclosporin A. Concomitant use of cyclosporin A will increase the blood levels of the statins; this has been shown to be most marked for pravastatin (7- to 23-fold increase) and lovastatin (4.2- to 7.8-fold increase), less so for simvastatin (2- to 6-fold increase) and cerivastatin (3- to 5-fold increase), and least for fluvastatin (1.3- to 1.9-fold increase). Data for atorvastatin are not available. The increased risk of myopathy seen with statins plus gemfibrozil may be expected because both of these agents alone can cause myopathy.

Teratogenicity in experimental animals has been demonstrated for lovastatin and fluvastatin but not for pravastatin or simvastatin. However, in view of the crucial role of cholesterol synthesis in proliferating cells, all of these drugs should be considered to be potentially harmful if used in pregnancy. An evaluation of 134 reports of exposure to lovastatin or simvastatin in the course of pregnancy showed a 4% incidence of congenital anomalies (Manson et al, 1996). This frequency is not higher than would be expected in the population at large. However, because of the limited number of reports of pregnancies with exposure to statins, these data allow us to conclude only that the frequency of congenital anomalies does not exceed 3 to 4 times the value expected in a population not exposed to statins.

In many different systems, each of these drugs has been shown not to be mutagenic. Therefore, a woman's exposure to these drugs prior to pregnancy should pose no risk to a fetus.

A generalized *eczematous rash* has been reported as a rare side effect of simvastatin. This appears to be caused by inhibition of cholesterol synthesis in the stratum corneum of the skin. Thus it seems possible that any of the statins could cause this problem.

The most common adverse effects of the statins are *dyspepsia, heartburn,* and *abdominal pain.* These occur in approximately 4% of persons treated with the statins. These six drugs do not differ substantially from one another in regard to the frequency of this side effect.

■ Drug Interactions

The influence of other drugs on the statins is summarized in Table 10.7. The interactions of greatest clinical significance are those involving bile acid sequestrants (reduced statin bioavailability with simultaneous administration) and with drugs shown to potentiate the myopathic potential of lovastatin (cyclosporin A, gemfibrozil, nicotinic acid, erythromycin, azole antifungals). Although data are lacking, these five drugs should also be considered to increase the myopathic potential of simvastatin, pravastatin, atorvastatin, fluvastatin, and cerivastatin.

Atorvastatin, cerivastatin, lovastatin, and simvastatin are metabolized by the cytochrome P450 (CYP) enzyme 3A4; fluvastatin and pravastatin do not rely on this enzyme for metabolism. Thus concomitant treatment with inhibitors of CYP 450 3A4 (eg, cyclosporin A, erythromycin, and azole antifungals) cause substantial increases in the plasma levels of atorvastatin, cerivastatin, lovastatin, and simvastatin. This pharmacokinetic interaction may underlie the increased risk of myopathy seen with these drugs. If so, cyclosporin A and other inhibitors of CYP 450 3A4 may confer less risk when they are used in combination with fluvastatin or provastatin rather than the other statins.

Small increases in prothrombin time have been shown when simvastatin was administered to patients treated with warfarin. Additionally, simvastatin has been shown to produce small increases in serum lev-

TABLE 10.7 — DRUG INTERACTIONS: INFLUENCE OF OTHER DRUGS ON HMG-COA REDUCTASE INHIBITORS

Statins	Bile Acid Sequestrants	Cyclosporine	Niacin, Gemfibrozil, Erythromycin, Azole Antifungals	Propranolol	Histamine$_2$-Blockers/ Omeprazole
Atorvastatin	↓ Bioavailability	↑ Myopathy risk	↑ Myopathy risk	??	No effect of cimetidine
Cerivastatin	↓ Bioavailability	↑ Myopathy risk	↑ Myopathy risk	??	No effect of cimetidine
Fluvastatin	↓ Bioavailability	↑ Bioavailability	? ↑ Myopathy risk	No effect	↑ Bioavailability
Lovastatin	??	↑ Bioavailability ↑ Myopathy risk	↑ Myopathy risk	Altered bioavail-ability (– 20%)	??
Pravastatin	↓ Bioavailability	↑ Bioavailability	? ↑ Myopathy risk; gemfibrozil – ↑ bioavailability	Altered bioavail-ability (– 20%)	No effect
Simvastatin	??	↑ Bioavailability ↑ Myopathy risk	? ↑ Myopathy risk	No effect	??

Abbreviation: HMG-CoA, 3-hydroxy-3-methyglutaryl coenzyme A.

els of digoxin. The other statins have not been shown to produce any significant impact on these parameters.

■ Cost

The cost of a year's treatment with different doses of the currently marketed statins is given in Table 10.8 as average wholesale price (April, 2000). The dose-response curve of the statins tends to plateau, as shown in Figure 10.1 for simvastatin. As the dose rises, an increase in the dose gives a less than proportional increase in LDL-c reduction. Because of the plateau of the dose-response curve, if cost is proportional to dose, these drugs will generally deliver the most LDL-c reduction per dollar when used in lower doses. In several cases (Table 10.8), the cost does not increase in proportion to dose. For example, simvastatin (Zocor) 80 mg/d is priced the same as simvastatin 20 mg/d and 40 mg/d. Here, higher doses are more cost-effective.

We should comment that cost and reduction of LDL cholesterol should not be the sole considerations in selecting one of these drugs. There is greater certainty of efficacy for clinical end points and of safety for those drugs which have seen successful use in a large-scale clinical trial. At this time, cerivastatin is the only statin not to have been used in a large-scale trial. Therefore, until cerivastatin has been successfully used in large clinical trials, this drug should be reserved for situations where it presents clear advantages and alternatives are unacceptable for reasons of efficacy or adverse effects.

As a final note, we should comment that costs can change and insurance considerations (eg, listing of items available on formulary for particular patients) can sometimes cause the cost to the patient to bear little relationship to the average wholesale cost. When decisions are being made on the basis of cost, it is important to be certain that the data used are current.

TABLE 10.8 — COST OF TREATMENT WITH LIPID-LOWERING MEDICATIONS	
Drug (Trade Name)	**Cost ($/y)**
*Bile Acid Sequestrants**	
Cholestyramine (LoCholest) – 8 g bid	1485
Cholestyramine (Prevalite) – 8 g bid	1240
Cholestyramine (Questran) – 8 g bid	1512
Cholestyramine generic (Major)	1135
Colestipol (Colestid) – 10 g bid	1446
Gemfibrozil (Lopid) – 600 mg bid	
Trade name	995
Generic (Major)	685
Fenofibrate (Tricor) – 200 mg/d	753
Niacin – 3 g/d	
Niacor	617
Niaspan	908
Slo-Niacin	210
Generic (Rugby)	60
HMG-CoA Reductase Inhibitors	
Atorvastatin (Lipitor)	
10 mg/d	686
20 mg/d	1060
40 mg/d	1277
80 mg/d	2553
Cerivastatin (Baycol)	
0.2 mg/d	482
0.3 mg/d	482
0.4 mg/d	482
Fluvastatin (Lescol)	
20 mg/d	458
40 mg/d	458
40 mg bid	916
Lovastatin (Mevacor)	
10 mg/d	482
20 mg/d	850

40 mg/d	1530
40 mg bid	3060
Pravastatin (Pravachol)	
10 mg/d	784
20 mg/d	830
40 mg/d	1364
Simvastatin (Zocor)	
5 mg/d	650
10 mg/d	796
20 mg/d	1389
40 mg/d	1389
80 mg/d	1389

Abbreviation: HMG-CoA, 3-hydroxy-3-methylglutaryl coenzyme A.

* Purchased as bulk powder.

Data taken from Red Book, April, 2000.

FIGURE 10.1 — SIMVASTATIN: DOSE-RESPONSE CURVE

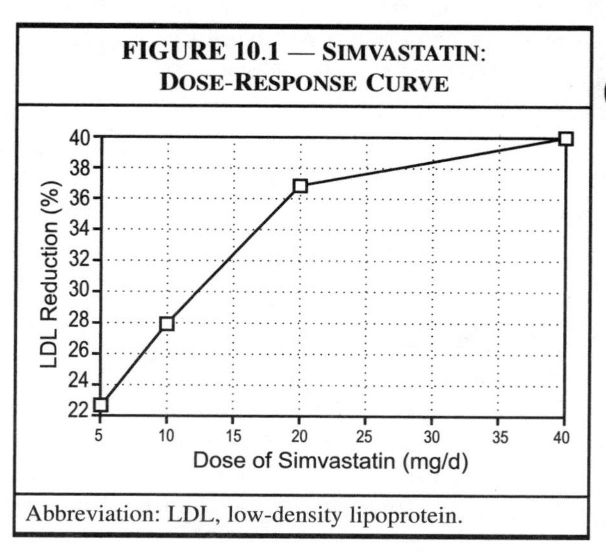

Abbreviation: LDL, low-density lipoprotein.

10

Bile Acid Sequestrants

The two drugs cholestyramine (Questran, Questran Light, LoCholest, LoCholest Light, and Prevalite) and colestipol (Colestid) have utility for reducing plasma levels of LDL-c. The bile acid sequestrants have been in use since the 1960s. Through most of the 1980s, they were the mainstay of pharmacologic therapy for reducing LDL-c. More recently, however, this role has been taken on by the HMG-CoA reductase inhibitors, which have improved tolerability and the potential for greater reductions in LDL-c. Currently, the bile acid sequestrants are most commonly used as adjunctive therapy when LDL-c reduction is not sufficient with a statin alone.

> ***Clinical Tip:*** A useful therapeutic maneuver is to add a small dose of bile acid sequestrant to boost the LDL-c lowering achieved with the HMG-CoA reductase inhibitors. A scoop of colestipol or cholestyramine resin mixed with psyllium powder and taken in the morning (always at least 60 to 90 minutes after taking any other medications) can significantly augment LDL-c lowering obtained with HMG-CoA reductase inhibitor therapy.

For these drugs, concern about systemic side effects is minimal because they are not absorbed from the gastrointestinal (GI) tract. Therefore, they may be particularly attractive for younger adults contemplating many decades of exposure to drug. The bile acid sequestrants are the only cholesterol-lowering drugs recommended by the National Cholesterol Education Program for routine use in children. They are also the only cholesterol-lowering drugs that can be used in pregnancy.

Both cholestyramine and colestipol are available as powders, which are suspended in liquid and then swallowed. Colestipol is also available in tablets of 1 g each.

■ Mechanism of Action

These drugs are anion-exchange resins. Their use initiates a complex set of events which result in increased LDL receptor activity and increased clearance of LDL from plasma. The bile acid sequestrants act within the intestine to bind bile acids, thereby interrupting the enterohepatic circulation of bile acids. The subsequent reduction in bile acid concentration within hepatocytes lifts the feedback inhibition of cholesterol-7-alpha-hydroxylase, the rate-limiting enzyme in bile acid synthesis. As cholesterol is converted to bile acid at an increased rate, the concentration of cholesterol within the hepatocyte falls.

Two homeostatic mechanisms are then activated to restore intracellular cholesterol concentrations. The first of these increases LDL receptor activity on hepatocyte cell membranes; this, in turn, *enhances the clearance of LDL from plasma.* This reduces LDL concentrations in plasma. The other homeostatic mechanism is an increase in the activity of HMG-CoA reductase (the rate-limiting enzyme of cholesterol biosynthesis) and a consequent increase in cholesterol synthesis in the hepatocyte.

■ Effects on Plasma Lipoprotein Levels

The best and largest set of data on the efficacy of the bile acid sequestrants comes from the Lipid Research Clinics Coronary Primary Prevention Trial (LRC-CPPT). In this carefully performed study, 3806 men were randomly assigned to treatment with cholestyramine or placebo. All participants were instructed to take cholestyramine in a dose of 12 g (3 packets) bid, but compliance varied. A linear relationship was seen between the consumption of cholestyramine and reduction in LDL-c (Table 10.9). Reduction in coronary risk corresponded to the changes in the dose of medication and to the reduction in LDL-c.

While the predominant effect of cholestyramine is a reduction of LDL-c, it also influences VLDL levels and HDL-c levels. Hypertriglyceridemia, reflecting increased secretion of VLDL, is a side effect of bile acid sequestrants. This effect may be quite marked in people with baseline elevations of triglyceride levels. Thus, hypertriglyceridemia mitigates against the use of a bile acid sequestrant.

It is important to recognize that the bile acid sequestrants are to be used only when hypercholesterolemia is due to an elevation of LDL-c. When the cholesterol elevation is due to increased amounts of VLDL (and the predominant lipoprotein abnormality is hypertriglyceridemia), the use of a bile acid sequestrant will be deleterious.

The bile acid sequestrants cause a small increase in HDL-c levels, generally 2% to 3%.

■ Adverse Effects

Because the bile acid sequestrants are not systemically absorbed, they cause a paucity of systemic side effects. This has been a major attraction of these agents. The predominant side effects of the bile acid sequestrants have been as follows:

- Constipation
- Bloating
- Heartburn.

These side effects are due to physical properties of these drugs. In the LRC-CPPT, during the initial year of treatment, constipation was reported by 39% of those assigned to cholestyramine but by only 10% of those assigned to a placebo. It appears that tolerance

TABLE 10.9 — RELATIONSHIP OF DOSE OF CHOLESTYRAMINE TO REDUCTION IN LDL-C AND IN CORONARY RISK IN THE LIPID RESEARCH CLINICS CORONARY PRIMARY PREVENTION TRIAL

Dose (g/d)	*n*	% Δ in LDL-c	% Δ in HDL-c	% Δ in TG	CHD Risk
0-4	294	– 6.6	+ 5.2	+ 10.7	– 10.9
4-8	145	– 8.7	+ 2.3	+ 12.7	
8-12	135	– 3.1	+ 5.5	+ 12.9	– 26.1
12-16	156	– 16.5	+ 6.0	+ 14.2	
16-20	205	– 20.9	+ 3.8	+ 15.5	– 39.3
20-24	965	– 28.3	+ 4.3	+ 17.1	

Abbreviations: LDL, low-density lipoprotein; *n*, number of patients; LDL-c, low-density lipoprotein cholesterol; HDL-c, high-density lipoprotein cholesterol; TG, triglyceride; CHD, coronary heart disease.

The Lipid Research Clinics Coronary Primary Prevention Trial results. II. *JAMA*. 1984;251:365-374.

10

does develop to the constipation occurring with the bile acid sequestrants; during the seventh year of the LRC-CPPT, constipation was reported by only 8% of those in the cholestyramine group.

When given in very large doses, cholestyramine can cause hyperchloremic acidosis.

As noted above, cholestyramine does increase plasma triglyceride levels, particularly in those patients with baseline hypertriglyceridemia.

■ Hints to Improve Compliance

Careful attention to measures designed to minimize the GI side effects of these medications is important in optimizing compliance. The initial dose should be small, 4 g/d for cholestyramine or 5 g/d for colestipol, and the dose should be increased gradually over 2 to 3 weeks to 4 to 8 g bid for cholestyramine or 5 to 10 g bid for colestipol. Patients should be forewarned about the possibility of constipation and should be advised to use a stool softener such as docusate sodium in doses up to 1200 mg/d. The addition of a bulk-forming laxative (eg, psyllium) also is useful, as is maintaining an adequate oral intake of fluids. When a bile acid sequestrant is being used, suspending it in prune juice can also be helpful in preventing constipation.

■ Use in Major Clinical Trials

The bile acid sequestrants have been used alone or in combination with other lipid-lowering medication in one major clinical trial with a clinical end point (LRC-CPPT) and in five major trials with angiographic end points, namely, the National Heart, Lung, and Blood Institute's (NHLBI) Type II Coronary Intervention Study, the Cholesterol Lowering Atherosclerosis Study (CLAS), the Familial Atherosclerosis Treatment Study (FATS), the St. Thomas Arteriosclerosis Regression Study (STARS), and the Lipoproteins in Coronary Atherosclerosis Study (LCAS) (see Chapter 7, *Clinical Trials*).

Drug Interactions

Since the bile acid sequestrants are anion-exchange resins, they have the potential to interfere with the absorption of concurrently administered anionic drugs. This interaction has been demonstrated for warfarin, l-thyroxine, hydrochlorothiazide, pravastatin, fluvastatin, and cerivastatin. Since most drugs (including lovastatin and simvastatin) have not been tested for an interaction with bile acid sequestrants, it would be prudent to avoid concurrent dosing with other medications: as a general rule, other medications should be administered at least 1 hour prior to or 4 hours after administration of a bile acid sequestrant.

Cost

The average wholesale price for 1 year's treatment with cholestyramine 8 g bid when purchased as the bulk powder is $1135 to $1512 (Table 10.8); for colestipol 10 g bid, the annual cost is $1446. When these agents are purchased in packets containing 4 g cholestyramine or 5 g colestipol or as the 1-g tablets of colestipol, the cost is greater.

Nicotinic Acid

Nicotinic acid (niacin) is the oldest known lipid-lowering agent. It has been used to treat hyperlipidemia for the past 40 years. It is a B vitamin. In doses considerably higher than those necessary to prevent a deficiency syndrome, it reduces plasma levels of VLDL and LDL, and it increases the levels of HDL. A range of troublesome side effects make nicotinic acid more difficult to use than some of the other medications discussed here.

Nicotinic acid is recommended as a first-line drug for treatment of hypertriglyceridemia and low HDL-c levels. When an elevated LDL-c level is associated with low HDL-c, nicotinic acid is an excellent choice. Additionally, in patients with high LDL-c levels, it can be used with excellent results in combination with a statin or with a bile acid sequestrant.

■ Mechanism of Action and Metabolism

Nicotinic acid's primary effect on lipoprotein metabolism is to reduce hepatic production of VLDL; this seems to result at least in part from reduced flux of fatty acids, a substrate for VLDL production, from adipose tissue to the liver. Nicotinic acid inhibits lipolysis in adipocytes. In clinical testing, the reduced VLDL levels are manifested as lower fasting triglyceride concentrations.

Since LDL is a metabolic product of VLDL, the reduced secretion of VLDL causes reduced LDL levels.

Nicotinic acid increases HDL-c levels more than any other lipid-altering medication. The increased HDL-c levels are due to retarded clearance of HDL; this, in turn, may be secondary to reduced plasma triglyceride levels. When triglyceride levels are elevated, there is increased exchange of triglyceride (in VLDL and chylomicrons) for cholesteryl esters (in HDL and LDL) in a process catalyzed by the cholesteryl ester transfer protein. The result of this process is an HDL particle depleted of cholesterol and enriched in triglyceride. The triglyceride that has been transferred into HDL can then be hydrolyzed by lipoprotein lipase (LPL). Thus the removal of HDL mass is stimulated by an elevated triglyceride level. Medications such as nicotinic acid, which have a primary effect of reducing VLDL levels, therefore may impede this process and secondarily increase HDL levels.

Nicotinic acid is rapidly and completely absorbed from the stomach and small intestine. Nearly 90% of an oral dose is found in the urine either as native drug or as metabolites.

■ Efficacy—Effects on Plasma Lipoproteins

Nicotinic acid in doses of 3 to 4.5 g/d typically:

- Reduces LDL-c levels by 20% to 25%
- Reduces triglyceride levels by 20% to 50%
- Increases HDL-c levels by 25% to 50%.

It is unique among the usual lipid-altering agents in having a substantial impact on Lp(a) levels; a dose of 3 g/d will reduce Lp(a) levels by an average of 30%. Sustained-release preparations of nicotinic acid cause less cutaneous flushing, but they tend to be less efficacious and also to have a higher incidence of GI side effects.

> ***Clinical Tip:*** Nicotinamide (niacinamide) is a nicotinic acid derivative capable of preventing the vitamin deficiency syndrome, but it is ineffective in altering lipoprotein levels.

10

■ Efficacy—Clinical and Atherosclerosis End Points

In clinical trials, nicotinic acid has reduced:

- Total mortality
- CHD mortality
- Nonfatal myocardial infarction (MI).

In angiographic studies, it has been shown to aid in limiting progression and enhancing regression of atherosclerotic lesions. (See Chapter 7, *Clinical Trials,* for details of the following studies involving nicotinic acid: Coronary Drug Project [CDP], CLAS, FATS, and Stockholm Ischemic Heart Disease Study.)

■ Adverse Effects

Bothersome side effects (Table 10.10) are unfortunately common with nicotinic acid. For this reason, approximately 30% of individuals are unable to tolerate this medication.

TABLE 10.10 — NICOTINIC ACID: ADVERSE EFFECTS
• Cutaneous flushing
• Dry skin, itching
• Acanthosis nigricans
• Gastritis
• Hepatitis
• Increased uric acid, gout
• Hyperglycemia
• Hypotension and syncope (uncommon)
• Atrial arrhythmias (uncommon)
• Toxic amblyopia (rare)

Cutaneous flushing occurs in all individuals treated with therapeutic doses of nicotinic acid. Tachyphylaxis occurs rapidly, and in most individuals flushing subsides or is greatly mitigated after 1 to 2 weeks. Flushing due to nicotinic acid is mediated by prostaglandins, and it can be eliminated or minimized by pretreatment with modest doses of aspirin or other prostaglandin inhibitors. Flushing is exacerbated when nicotinic acid is consumed with hot beverages, which may increase its rate of absorption.

Gastritis and *peptic ulcer disease* are the most common reasons for inability to tolerate nicotinic acid. They occur more frequently when sustained-release preparations of the drug are used or when nicotinic acid is taken without food.

Hepatitis occurs in approximately 3% of individuals treated with nicotinic acid. When this occurs with nicotinic acid, it is generally accompanied by nausea and abdominal pain (in contrast to the transaminase elevation seen with the statins, which is usually not associated with symptoms). Nicotinic acid–associated hepatitis is dose-related, and individuals who develop it are often able to tolerate a reduced dose of the drug. It occurs much more commonly with sustained-release preparations of nicotinic acid. It has been suggested that a recently developed brand of sustained-release nicotinic acid (Niaspan) may be better tolerated and less hepatotoxic than previously available preparations. However, recent studies investigating this are best regarded as inconclusive due to either: (Knopp et al, 1998; Goldberg et al, 2000)

- Limited duration of follow-up
- Small sample size
- Use of relatively low doses of the sustained-release preparation of nicotinic acid.

Cutaneous side effects are common with nicotinic acid. These include:

- Dry skin
- Ichthyosis
- Acanthosis nigricans.

Increases in uric acid level are the rule when nicotinic acid is used, and attacks of acute *gout* can occur. In these cases, ongoing treatment with colchicine and/or allopurinol can generally allow the continued use of nicotinic acid.

Hyperglycemia occasionally occurs in patients taking nicotinic acid. Patients who are borderline diabetic can become overtly diabetic. Diabetics who are taking medication may require larger doses of their medications or more potent medications if they undergo treatment with nicotinic acid.

■ Hints to Improve Compliance

A variety of measures can improve the tolerability of nicotinic acid and improve compliance (Table

10.11). In order to minimize flushing, the initial dose should be very small, eg, 50 mg tid. The dose can be doubled every 3 days until a total daily dose of 1500 to 3000 mg (administered in divided doses) is attained. This dose level should be maintained for at least 1 month before assessing its impact on lipoprotein levels.

TABLE 10.11 — NICOTINIC ACID: MEASURES TO IMPROVE COMPLIANCE AND MINIMIZE SIDE EFFECTS

- Begin with low dose and increase gradually.
- Take aspirin or other prostaglandin inhibitor to minimize flushing.
- Do not take without food.
- Do not take with hot beverage.
- Begin with immediate-release preparation of drug; switch to sustained-release preparation only if flushing is intolerable despite use of prostaglandin inhibitors.

Nicotinic acid should be administered with food in order to minimize flushing and the likelihood of abdominal pain. By avoiding concurrent consumption of hot beverages, absorption can be slowed and flushing minimized.

Aspirin or another prostaglandin inhibitor taken 30 minutes prior to a dose of nicotinic acid can diminish or abolish flushing. It is often most effective to take an aspirin tablet (or ibuprofen 200 mg) at the onset of a meal and nicotinic acid at the end of a meal; even smaller doses of prostaglandin inhibitors are often fully effective.

When compliance is harmed by intense cutaneous flushing or the 3-times-daily dosing frequency of immediate-release preparations of nicotinic acid, sustained-release preparations may be considered. However, sustained-release preparations of nicotinic acid may be associated with a greater frequency of abdominal pain or hepatitis.

Cost

Generic nicotinic acid is the least expensive of the lipid-lowering drugs (Table 10.8). Treatment typically may be as low as $60 per year (average wholesale price). Brand preparations of nicotinic acid can be much more costly.

Fibric Acid Derivatives

Three drugs in the fibric acid class are available in the United States:

- Gemfibrozil (Lopid)
- Clofibrate (Atromid-S)
- Fenofibrate (Tricor).

In other countries, bezafibrate is also available (Table 10.12).

TABLE 10.12 — FIBRIC ACID DRUGS: USUAL DOSING

Drug	Dosage
Bezafibrate	600 mg bid
Clofibrate	1000 mg bid
Fenofibrate	67-200 mg/d
Gemfibrozil	600 mg bid

Efficacy and Mechanism of Action

The major effect of the fibric acid derivatives is to reduce levels of VLDL by increasing its intravascular metabolism through increased LPL activity. Additionally, there may be some up-regulation of the LDL receptor (Goto et al, 1997). In routine laboratory testing, this is manifested as a reduction in fasting plasma triglyceride levels, which typically fall 20% to 70%. The effect of the fibric acid derivatives on LDL-c is modest. LDL-c usually falls by about

10%; however, in hypertriglyceridemic individuals, these drugs may cause LDL-c levels to *rise*. HDL-c levels typically rise. In the Helsinki Heart Trial, treatment with gemfibrozil, 600 mg bid, initially caused a 10% reduction in LDL-c, a 10% increase in HDL-c, and a 43% reduction in triglycerides.

The fibric acid derivatives have been shown to limit the insulin-stimulated increase in plasminogen activator inhibitor type 1 (Nordt et al, 1997). This may lead to improved fibrinolytic activity in hyperinsulinemic patients.

■ Metabolism and Excretion

The fibric acid derivatives undergo conjugation to glucuronides in the liver and renal excretion. Therefore, clearance of the native drugs or their metabolites will be retarded in patients with impaired liver or kidney function. Thus the doses of these drugs must be reduced in such patients.

■ Implications of Clinical Trials

(See Chapter 7, *Clinical Trials*)

Six major clinical trials have utilized the fibric acid drugs:

- The Coronary Drug Project (CDP) (clofibrate)
- The World Health Organization (WHO) Clofibrate Trial
- The Helsinki Heart Trial (HHT) (gemfibrozil)
- The Bezafibrate Coronary Atherosclerosis Intervention Trial (BECAIT)
- The Lopid Coronary Angiography Trial (LOCAT)
- The Veterans Affairs HDL Intervention Trial (VA-HIT).

While reduced rates of CHD events were seen in the CDP, the WHO trial, and in the HHT, the results of these three large studies have raised concerns about the safety of these drugs.

In the CDP, clofibrate afforded no benefit with regard to total mortality (the study's primary end point), nonfatal MI, or cause-specific mortality. There was an increased frequency of angina pectoris, thrombophlebitis, and arrhythmias other than atrial fibrillation. In the WHO study, although there was a reduction in nonfatal CHD events, total mortality and non-CHD mortality were increased in the clofibrate group. Finally, in the HHT, during extended follow-up, all-cause mortality was slightly higher in the men of the original gemfibrozil group than among those originally assigned to placebos. (The BECAIT and LOCAT trials were studies with angiographic end points, involving much smaller numbers of patients and with many fewer clinical events.)

A meta-analysis of seven clinical trials of fibric acid drugs showed a significantly increased risk of non-CHD mortality (odds ratio 1.29) and of stroke (odds ratio 1.91) with a nonsignificant trend toward increased cases of cancer (Rossouw, 1995).

However, in the VA-HIT (with 2531 participants), the gemfibrozil treatment was associated with a significant 22% reduction in the primary end point (nonfatal MI or CHD death) without evidence of significant adverse effects. In fact, there was nonsignificant reduction in overall mortality and in the incidence of cancer in the gemfibrozil group.

In light of these findings, how are these drugs to be used? A restrospective analysis of data from the HHT showed the greatest benefit from gemfibrozil was derived by a specific high-risk group of patients: those with both a high ratio of LDL-c/HDL-c (>5.0) and a triglyceride level >200 mg/dL. In this group, gemfibrozil was associated with a 71% reduction in CHD risk. The VA-HIT, however, demonstrated benefits in its population (low HDL, normal LDL, triglyceride ≤300 mg/dL) regardless of baseline levels of HDL, LDL, or triglyceride. In view of these discrepant results, it seems reasonable at this time to consider fibrates for patients with low HDL levels and for patients with hypertriglyceridemia who also have a high ratio of LDL-c/HDL-c. Additionally, fibrates are the drugs of choice for the rare patients with dysbetalipoproteinemia (type III hyperlipoproteinemia).

■ Adverse Effects and Drug Interactions

The major side effects of the fibric acid drugs are listed in Table 10.13. These drugs increase the lithogenicity of bile. For this reason, they predispose to *gallstone* disease.

TABLE 10.13 — SIDE EFFECTS OF FIBRIC ACID DRUGS
Common
• Gallstone disease
• Dyspepsia and abdominal pain
• Decreased libido
• ?? Increased noncoronary heart disease mortality
Rare
• Myositis
• Ventricular arrhythmia
• Leukopenia

The fibric acid drugs are strongly bound to albumin, and they *displace warfarin* from albumin. Therefore, patients taking warfarin must be closely monitored when treatment with a fibric acid derivative is initiated or discontinued.

Concurrent use of a fibric acid derivative and an HMG-CoA reductase inhibitor may markedly increase the likelihood of *myopathy* (discussed earlier).

Cost

The average wholesale price for the Lopid brand of gemfibrozil is $995 for 1 year's treatment at a dose of 600 mg bid. Generic gemfibrozil is listed with an average wholesale price as low as $685 per year. The average wholesale price for 1 year of treatment with micronized fenofibrate (Tricor) is $753.

Fish Oils

The somatic fish oils can be very useful in treatment of severe hypertriglyceridemias (triglycerides >1000 mg/dL). With high doses of fish oils, fasting plasma triglyceride levels often fall by 75%. Typical doses of somatic fish oils are 2 g to 6 g tid.

Mechanism of Action

Highly polyunsaturated n-3 fatty acids (eicosapentaenoic acid [EPA; 20:5] and docosahexaenoic acid [DHA; 22:6]), when given in large quantities, have a dramatic inhibitory effect on the secretion of VLDL due to increased intracellular degradation of apo B-100. These fatty acids are found primarily in the somatic oils of fish. When given in large quantities, they cause a marked fall in the VLDL plasma level. Plasma LDL-c levels also fall when these oils are fed to normolipidemic volunteers; this reduction in LDL-c levels is similar to that seen with omega-6 polyunsaturated fatty acids. However, in hypertriglyceridemic patients, these oils along with reducing triglyceride levels often *raise* LDL-c levels; this increase in LDL-c is similar to that seen when hypertriglyceridemic patients are treated with the fibric acid drugs (see Chapter 8, *Dietary Therapy for Hyperlipidemia*.)

Adverse Effects

The high doses of somatic fish oils used for treating hypertriglyceridemia carry a burden of dietary calories. A dose of 6 g tid is approximately 160 kcal/d. Concerns had been raised that fish oils may occasionally lead to episodes of bleeding because they interfere with platelet function; this has not been noted to occur in the various clinical trials of fish oils.

Combination Drug Therapy

There are five reasons for utilizing combinations of lipid-altering medications. These are summarized in Table 10.14.

TABLE 10.14 — REASONS FOR USING COMBINATIONS OF LIPID-ALTERING MEDICATIONS
• Maximize reduction of low-density lipoprotein cholesterol (LDL-c)
• Maximize reduction of very low-density lipoprotein
• Minimize side effects by using smaller doses of drugs
• Allow use of bile acid sequestrants in hypertriglyceridemic patients with elevated LDL-c
• Treat increased LDL-c found consequent to treatment of high triglycerides with fibric acid derivative

Therapy with a combined drug regimen should be considered only after diet and a single drug have been found to be insufficient. Combination therapy should be approached in step-wise fashion, adding one drug at a time. Plasma lipoprotein levels and possible drug side effects should be assessed 4 to 8 weeks after a new drug has been added. The addition of a third drug should be considered only after the effects of two drugs have been measured at least twice at intervals of no less than 4 to 6 weeks. If a newly added drug does not cause a beneficial response, that drug should be discontinued.

■ Combination Therapy to Maximize Reduction of LDL-c

(Table 10.15)

Since monotherapy for reducing LDL-c will usually be with a statin, most two-drug regimens will contain one of these drugs. The most effective two-drug regimens for reducing LDL-c are shown in Table 10.15. The greatest reduction of LDL-c has been found with the three-drug regimen of a bile acid sequestrant, nicotinic acid, and a statin. Regimens containing nicotinic acid have the benefits of producing the greatest increases of HDL-c and also substantial reductions in triglyceride levels.

When a statin is paired with a bile acid sequestrant, LDL-c reduction is synergistic for two reasons:

- These drugs increase LDL-c receptor activity by slightly different mechanisms
- The statin prevents increased cholesterol biosynthesis, which would occur with bile acid sequestrant monotherapy.

> ***Clinical Tip:*** Often, the use of small doses of the second agent (such as nicotinic acid, 500 mg tid, or cholestyramine, 4 g once or twice daily) can give considerable additional reduction in LDL-c concentration (Table 10.15).

It should be remembered that the combination of nicotinic acid with a statin may cause a small increase in the risk of myopathy. This possibility may warrant additional attentiveness to the occurrence of this side effect, but it is not a reason to avoid this very effective combination of medications.

■ Combination Therapy to Maximize Reduction in VLDL Levels

In markedly hypertriglyceridemic patients, sometimes the use of a single agent does not provide suffi-

TABLE 10.15 — COMBINATION REGIMENS FOR REDUCING LOW-DENSITY LIPOPROTEIN CHOLESTEROL

Drug Combinations	Usual % Change in LDL-c	Usual % Change in HDL-c
Statin + bile acid sequestrant	– 50	+ 10-15
Statin + nicotinic acid	– 50	+ 25-50
Nicotinic acid + bile acid sequestrant	– 35	+ 25-50
Statin + bile acid sequestrant + nicotinic acid	– 66	+ 25-50

Abbreviations: LDL, low-density lipoprotein; LDL-c, low-density lipoprotein cholesterol; HDL-c, high-density lipoprotein cholesterol.

cient reduction of triglyceride levels. In these patients, the combination of two or three triglyceride-reducing agents (nicotinic acid, a fibric acid derivative, or somatic fish oils) acting by different mechanisms can provide additional reduction of triglyceride levels.

■ Combination Therapy to Minimize Doses and Side Effects

The side effects of LDL-reducing drugs tend to be dose-dependent. Combination regimens using small doses of bile acid sequestrants (colestipol, 5 to 10 g/d, or cholestyramine, 4 to 8 g/d), nicotinic acid (500 mg bid or tid), and/or statins can often be well tolerated when higher doses are not, and in most cases they can yield a sufficient reduction of LDL-c.

■ Combination Therapy to Allow Use of a Bile Acid Sequestrant in Hypertriglyceridemic Patients With Concurrent Elevation of LDL-c

Bile acid sequestrants typically increase the secretion of VLDL, and this is manifested as an increase in fasting plasma triglyceride levels. When the baseline triglyceride level is elevated, this phenomenon can preclude the use of a bile acid sequestrant as the sole agent for treatment of concurrently elevated LDL-c levels. However, the addition of a triglyceride-lowering medication (nicotinic acid, fibric acid derivative, or fish oil) can allow us to use a bile acid sequestrant to reduce LDL-c levels in these patients. The best of these combinations is often nicotinic acid plus a bile acid sequestrant because both of these medications are potent in reducing LDL-c.

■ Combination Therapy to Treat the Increased LDL-c Levels Occurring Consequent to Use of a Fibric Acid Derivative in Hypertriglyceridemic Patients

When a fibric acid derivative is used in patients with hypertriglyceridemia, there is often a secondary *increase* in LDL-c levels. Thus, even though triglyceride levels fall, the patient's condition may be worse with treatment (because of elevated LDL-c) than without it. However, the addition of a drug to reduce LDL-c levels (nicotinic acid, a bile acid sequestrant, or a statin) can allow hypertriglyceridemia to be treated with a fibric acid derivative. The clinician must be mindful of the increased risk of myopathy that occurs with the combination of a statin and a fibric acid derivative; this combination, although occasionally useful, does warrant caution.

SELECTED READINGS

Bile Acid Sequestrants

The Lipid Research Clinics Coronary Primary Prevention Trial results. I. Reduction in incidence of coronary heart disease. *JAMA*. 1984;251:351-364.

The Lipid Research Clinics Coronary Primary Prevention Trial results. II. The relationship of reduction in incidence of coronary heart disease to cholesterol lowering. *JAMA*. 1984;251:365-374.

Combination Drug Therapy

La Rosa JC. Combinations of drugs in lipid-lowering therapy. *Am J Med*. 1994;96:399-400. Editorial.

Leitersdorf E, Muratti EN, Eliav O, et al. Efficacy and safety of a combination fluvastatin-bezafibrate treatment for familial hypercholesterolemia: comparative analysis with a fluvastatin-cholestyramine combination. *Am J Med*. 1994;96:401-407.

Fibric Acid Derivatives

Frick MH, Elo O, Haapa K, et al. Helsinki Heart Study: primary-prevention trial with gemfibrozil in middle-aged men with dyslipidemia. Safety of treatment, changes in risk factors, and incidence of coronary heart disease. *N Engl J Med*. 1987;317: 1237-1245.

Goto D, Okimoto T, Ono M, et al. Upregulation of low density lipoprotein receptor by gemfibrozil, a hypolipidemic agent, in human hepatoma cells through stabilization of mRNA transcripts. *Arterioscler Thromb Vasc Biol.* 1997;17:2707-2712.

Manninen V, Tenkanen L, Koskinen P, et al. Joint effects of serum triglyceride and LDL cholesterol and HDL cholesterol concentrations on coronary heart disease risk in the Helsinki Heart Study: implications for treatment. *Circulation.* 1992; 85:37-45.

Nordt TK, Kornas K, Peter K, et al. Attenuation by gemfibrozil of expression of plasminogen activator inhibitor type 1 induced by insulin and its precursors. *Circulation.* 1997;95:677-683.

Rossouw JE. Non-CHD mortality in cholesterol-lowering trials. *Cardiovasc Risk Factors.* 1995;5:181-188.

Rubins HB, Robins SJ, Collins D, et al. Gemfibrozil for the secondary prevention of coronary heart disease in men with low levels of high-density lipoprotein cholesterol. Veterans Affairs High-Density Lipoprotein Cholesterol Intervention Trial Study Group. *N Engl J Med.* 1999;341:410-418.

HMG-CoA Reductase Inhibitors

Akhlaghi F, McLachlan AJ, Keogh AM, Brown KF. Effect of simvastatin on cyclosporine unbound fraction and apparent blood clearance in heart transplant recipients. *Br J Clin Pharmacol.* 1997;44:537-542.

Bakker-Arkema RG, Davidson MH, Goldstein RJ, et al. Efficacy and safety of a new HMG-CoA reductase inhibitor, atorvastatin, in patients with hypertriglyceridemia. *JAMA.* 1996;275:128-133.

Bertolini S, Bon GB, Campbell LM, et al. Efficacy and safety of atorvastatin compared to pravastatin in patients with hypercholesterolemia. *Atherosclerosis.* 1997;130:191-197.

Blum CB. Comparison of properties of four inhibitors of 3-hydroxy-3-methylglutaryl-coenzyme A reductase [published erratum appears in *Am J Cardiol.* 1994;74:639]. *Am J Cardiol.* 1994;73:3D-11D.

Bradford RH, Shear CL, Chremos AN, et al. Expanded Clinical Evaluation of Lovastatin (EXCEL) study results. I. Efficacy in modifying plasma lipoproteins and adverse event profile in 8245 patients with moderate hypercholesterolemia. *Arch Intern Med.* 1991;151:43-49.

Cilla DD Jr, Gibson DM, Whitfield LR, Sedman AJ. Pharmacodynamic effects and pharmacokinetics of atorvastatin after administration to normocholesterolemic subjects in the morning and evening. *J Clin Pharmacol.* 1996;36:604-609.

Cilla DD Jr, Whitfield LR, Gibson DM, Sedman AJ, Posvar EL. Multiple-dose pharmacokinetics, pharmacodynamics, and safety of atorvastatin, an inhibitor of HMG-CoA reductase, in healthy subjects. *Clin Pharmacol Ther.* 1996;60:687-695.

Crouse JR 3rd, Frohlich J, Ose L, Mercuri M, Tobert JA. Effects of high doses of simvastatin and atorvastatin on high-density lipoprotein cholesterol and apolipoprotein A-I. *Am J Cardiol.* 1999;83:1476-1477, A7.

Cummings SR, Bauer DC. Do statins prevent both cardiovascular disease *and* fracture? *JAMA.* 2000;283:3255-3257.

Dart A, Jerums G, Nicholson G, et al. A multicenter, double-blind, one-year study comparing safety and efficacy of atorvastatin versus simvastatin in patients with hypercholesterolemia. *Am J Cardiol.* 1997;80:39-44.

Davidson M, McKenney J, Stein E, et al. Comparison of one-year efficacy and safety of atorvastatin versus lovastatin in primary hypercholesterolemia. Atorvastatin Study Group I. *Am J Cardiol.* 1997;79:1475-1481.

Douste-Blazy P, Ribeiro VG, Seed M and the European Study Group. Comparative study of the efficacy and tolerability of simvastatin and pravastatin in patients with primary hypercholesterolemia. *Drug Invest.* 1993;6:353-361.

Endo A. The discovery and development of HMG-CoA reductase inhibitors. *J Lipid Res.* 1992;33:1569-1582.

Farmer JA, Washington LC, Jones PH, Shapiro DR, Gotton AM Jr, Mantell G. Comparative effects of simvastatin and lovastatin in patients with hypercholesterolemia. The Simvastatin and Lovastatin Multicenter Study Participants. *Clin Ther.* 1992;14:708-717.

Grundy SM. Consensus statement: Role of therapy with "statins" in patients with hypertriglyceridemia. *Am J Cardiol.* 1998;81:1B-6B.

Heinonen TM, Stein E, Weiss SR, et al. The lipid-lowering effects of atorvastatin, a new HMG-CoA reductase inhibitor: results of a randomized, double-masked study. *Clin Ther.* 1996;18:853-863.

Jones P, Kafonek S, Laurora I, Hunninghake D. Comparative dose efficacy study of atorvastatin versus simvastatin, pravastatin, lovastatin, and fluvastatin in patients with hypercholesterolemia (the CURVES study) [published erratum appears in *Am J Cardiol.* 1998;82:128]. *Am J Cardiol.* 1998;81:582-587.

Kobashigawa JA, Katznelson S, Laks H, et al. Effect of pravastatin on outcomes after cardiac transplantation. *N Engl J Med.* 1995;333:621-627.

Lambrecht LJ, Malini PL. Efficacy and tolerability of simvastatin 20 mg vs pravastatin 20 mg in patients with primary hypercholesterolemia. European Study Group. *Acta Cardiol.* 1993;48:541-554.

Laufs U, Liao JK. Post-transcriptional regulation of endothelial nitric oxide synthase mRNA stability by Rho GTPase. *J Biol Chem.* 1998;273:24266-24271.

Laufs U, Marra D, Node K, Liao JK. 3-Hydroxy-e-methylglutaryl-CoA reductase inhibitors attenuate vascular smooth muscle proliferation by preventing rho GTPase-induced down-regulation of p27(Kip1). *J Biol Chem.* 1999;274:21926-21931.

Manson JM, Freyssinges C, Ducrocq MB, Stephenson WP. Postmarketing surveillance of lovastatin and simvastatin exposure during pregnancy. *Reprod Toxicol.* 1996;10:439-446.

Meier CR, Schlienger RG, Kraenzlin ME, Schlegel B, Jick H. HMG-CoA reductase inhibitors and the risk of fractures. *JAMA.* 2000;283:3205-3210.

Muck W, Ritter W, Ochmann K, et al. Absolute and relative bioavailability of the HMG-CoA reductase inhibitor cerivastatin. *Int J Clin Pharmacol Ther.* 1997;35:255-260.

Naoumova RP, Dunn S, Rallidis L, et al. Prolonged inhibition of cholesterol synthesis explains the efficacy of atorvastatin. *J Lipid Res.* 1997;38:1496-1500.

Nawrocki JW, Weiss SR, Davidson MH, et al. Reduction of LDL cholesterol by 25% to 60% in patients with primary hypercholesterolemia by atorvastatin, a new HMG-CoA reductase inhibitor. *Arterioscler Thromb Vasc Biol.* 1995;15:678-682.

Peters TK, Muratti EN, Mehra M. Efficacy and safety of fluvastatin in women with primary hypercholesterolaemia. *Drugs.* 1994;47(suppl 2):64-72.

Stein E. Cerivastatin in primary hyperlipidemia: a multicenter analysis of efficacy and safety. *Am J Cardiol.* 1998;82:40J-46J.

Stein EA, Davidson MH, Dobs AS, et al. Efficacy and safety of simvastatin 80 mg/day in hypercholesterolemic patients. The Expanded Dose Simvastatin U.S. Study Group. *Am J Cardiol.* 1998;82:311-316.

Stein EA, Lane M, Laskarzewski P. Comparison of statins in hypertriglyceridemia. *Am J Cardiol.* 1998;81:66B-69B.

Steinhagen-Thiessen E. Comparative efficacy and tolerability of 5 and 10 mg simvastatin and 10 mg pravastatin in moderate primary hypercholesterolemia. Simvastatin Pravastatin European Study Group. *Cardiology*. 1994;85:244-254.

The Lovastatin Pravastatin Study Group. A multicenter comparative trial of lovastatin and pravastatin in the treatment of hypercholesterolemia. *Am J Cardiol*. 1993;71:810-815.

Wang PS, Solomon DH, Mogun H, Avorn J. HMG-CoA reductase inhibitors and the risk of hip fractures in elderly patients. *JAMA*. 2000;283:3211-3216.

Wenke K, Meiser B, Thiery J, et al. Simvastatin reduces graft vessel disease and mortality after heart transplantation: a four-year randomized trial. *Circulation*. 1997;96:1398-1402.

Nicotinic Acid

Canner PL, Berge KG, Wenger NK, et al. Fifteen year mortality in Coronary Drug Project patients: long-term benefit with niacin. *J Am Coll Cardiol*. 1986;8:1245-1255.

Goldberg A, Alagona P, Capuzzi DM, et al. Mutiple-dose efficacy and safety of an extended-release form of niacin in the management of hyperlipidemia. *Am J Cardiol*. 2000;85:1100-1105.

Knopp RH, Alagona P, Davidson M, et al. Equivalent efficacy of a time-release form of niacin (Niaspan) given once-a-night versus plain niacin in the management of hyperlipidemia. *Metabolism*. 1998;47:1097-1104.

Knopp RH, Ginsberg J, Albers JJ, et al. Contrasting effects of unmodified and time-release forms of niacin on lipoproteins in hyperlipidemic subjects: clues to mechanism of action of niacin. *Metabolism*. 1985;34:642-650.

The Coronary Drug Project Research Group. Clofibrate and niacin in coronary heart disease. *JAMA*. 1975;231:360-381.

Somatic Fish Oils

Wang H, Chen X, Fisher EA. N-3 fatty acids stimulate intracellular degradation of apoprotein B in rat hepatocytes. *J Clin Invest*. 1993;91:1380-1389.

11 Special Populations

Children

Since the diagnosis and treatment of children impact upon their parents and grandparents, a brief review of the National Cholesterol Education Program (NCEP) Report of the Expert Panel on Blood Cholesterol Levels in Children and Adolescents is presented. The panel recommended both a population approach and an individualized approach in dealing with the problem of high blood cholesterol in children and adolescents. They suggested that for all healthy children and adolescents over the age of 2 years, a nutritious, balanced Step I diet with adequate energy (calories) to support growth and development should be consumed.

> ***Clinical Tip:*** The Step I diet is not intended for infants from birth to 2 years, whose fast growth requires a higher percentage of calories from fat.

In addition, the panel made specific recommendations about identifying children and adolescents who were likely to have an increased risk of coronary heart disease (CHD). In the context of regular pediatric health care, children and adolescents with a family history of premature cardiovascular disease or at least one parent with high blood cholesterol should be selectively screened. Universal screening was not recommended, however. In brief, the recommendations included:

- Screening with a nonfasting blood cholesterol if one parent has high cholesterol

- Performing a fasting lipoprotein profile on a child whose cholesterol is 200 mg/dL or greater
- Screening with fasting lipoprotein profile if there is a documented history of premature CHD in a parent or grandparent (<55 years of age in males and <65 years of age in females).

> ***Clinical Tip:*** Measuring blood lipids in adult relatives of hypercholesterolemic children is a good way to find cases of high cholesterol that may benefit from clinical evaluation.

> ***Clinical Tip:*** Parents often ask if their adopted children need lipid testing. It is advised that they do get a screening cholesterol to be sure that they are not heterozygous for familial hypercholesterolemia.

Table 11.1 describes the classification of acceptable, borderline, and high values for total cholesterol and low-density lipoprotein cholesterol (LDL-c) as suggested by the Expert Panel. The goals of treatment are given in Table 11.2. The panel recommends the Step I diet as the initial dietary therapy. The Step II diet is recommended if careful adherence for at least 3 months has failed to achieve the minimal goals of therapy.

Diets lower in fat and cholesterol for children do not appear to affect growth and development, although improvements in lipids were not seen in one study of third grade students followed for 3 years (Luepker et al, 1996). The Dietary Intervention Study in Children (DISC) provided much-needed data about the safety and efficacy of a Step II type diet in children. This 3-year dietary intervention enrolled 362 prepubertal boys and 301 girls aged 8 to 10 with LDL-c levels in the 80th to 98th percentiles for age and sex. The intervention group was given a diet which provided 28%

TABLE 11.1 — CLASSIFICATION OF TOTAL CHOLESTEROL AND LDL-C

Category	Blood Cholesterol	LDL-c
Acceptable	<170 mg/dL	<110 mg/dL
Borderline	170-199 mg/dL	110-129 mg/dL
High	≥200 mg/dL	≥130 mg/dL

Abbreviation: LDL-c, low-density lipoprotein cholesterol.

Report of the Expert Panel on Blood Cholesterol Levels in Children and Adolescents. NIH Publication No. 91-2732. September, 1991.

TABLE 11.2 — GOALS OF THERAPY IN CHILDREN AND ADOLESCENTS

Category	Treatment Goals
If borderline LDL-c	Lower to <110 mg/dL
If high LDL-c	Lower to <130 mg/dL as a minimal goal; ideally ≤110 mg/dL

Abbreviation: LDL-c, low-density lipoprotein cholesterol.

Report of the Expert Panel on Blood Cholesterol Levels in Children and Adolescents. NIH Publication No. 91-2732. September, 1991.

of energy from total fat, <8% of energy as saturated fat, and less than 75 mg/1000 kcal as dietary cholesterol, which significantly lowered LDL-c levels. Importantly, they maintained adequate growth, iron stores, nutritional adequacy, and psychological well-being. An ambitious attempt to look at a cholesterol-lowering diet (reduced in saturated fat and dietary cholesterol) in 7-month-old children did show a small significant difference in non–high-density lipoprotein cholesterol (HDL-c) in boys but not in girls, with no adverse effects on growth (Niinikoski et al, 1996).

The advantage, however, of restricting fat and cholesterol in such young children is not proven and children under age 2 should not be given diets restricted in fat and cholesterol.

> ***Clinical Tip***: These data support the use of diet as the initial therapy for children and adolescents with high blood cholesterol. In addition, DISC suggests that low-fat choices should be recommended for places where children eat, eg, school cafeterias, fast-food restaurants, and homes.

The NCEP panel was conservative in its recommendations about drug therapy in children. It recommended consideration of drug therapy in children who are 10 years of age and older *only* if an adequate trial of diet (6 months to 1 year in duration) has occurred and the LDL-c remains ≥190 mg/dL *or* if the LDL-c is ≥160 mg/dL and a positive family history of premature CHD is present *or* if two or more other risk factors are still present in the child or adolescent (low HDL-c of <5 mg/dL, cigarette smoking, high blood pressure, obesity, or diabetes) after vigorous attempts are made to control or eliminate these risk factors.

Since the near-term risk of CHD in dyslipidemic children is low except in rare cases of genetic lipoprotein abnormalities, the threshold for drug treatment should be high. Children with familial hypercholesterolemia (FH) should be considered for lipid-lowering therapy based on their LDL-c level, presence of familial premature CHD, and other risk indicators for CHD, including male gender and Lp(a) elevations. Smoking should be strictly forbidden. Drug therapy with a resin (see Chapter 10, *The Lipid-Lowering Drugs*) and niacin has been preferred for safety reasons, although compliance is not good. Lovastatin has been given in a randomized, placebo-controlled trial in boys 10 to 17 years of age in outpatient pediatric clinics in the United States and Finland and found to be both safe and effective (Stein et al, 1999). Growth and sexual maturation assessed by Tanner staging and testicular volume were not significantly different between the lovastatin and placebo groups at 24 and 48 weeks. There were no significant changes in serum hormone levels or biochemical parameters of nutrition as well. We agree with the accompanying editorial that stated "Because clinical events are rare with this disorder during childhood and early adulthood, other measures are needed to assess the impact of treatment strategies initiated in childhood. Noninvasive methods of assessing atherosclerosis severity at different sites may allow future trials to address these issues without requiring unduly large numbers of children or excessively prolonged follow-up. Until such trials can be done, it seems productive to identify and treat heterozygous FH aggressively in adolescent boys, given the significant morbidity and mortality from CHD that occur as early as the third decade of life. The appropriate strategy is less certain for women with heterozygous FH, given their relatively delayed onset of clinical CHD sequelae and the paucity of safety data pertaining to conception and pregnancy" (Rifkind et al, 1999).

Clinical Tip: Counseling against cigarette smoking and endorsing regular exercise and diet to prevent obesity are among the most important things a pediatrician can do to reduce the risk of future CHD. For the child with very high levels of LDL-c despite a trial of diet, and especially if there is a family history of premature CHD and/or familial hypercholesterolemia, referral to a lipid specialist may be prudent to take advantage of the greater experience in treating such individuals.

One of the most informative among the many studies which have provided useful insights into lipid levels and cardiovascular risk in childhood is the Bogalusa Heart Study. Dr. Berenson and his colleagues noted in their 15-year study of a biracial community that high levels of LDL-c in childhood can predict adult dyslipidemia. Using stepwise multiple regression analysis, they showed that the childhood level of LDL-c followed by the change in body mass index (weight over height squared as a measure of obesity) were the most helpful in explaining the variability seen in total cholesterol, triglycerides, HDL-c, and LDL-c. They further noted in the children in Bogalusa that clustering of multiple risk factors also tracks from childhood to adulthood and the correlation was, in fact, stronger than for individual factors.

> ***Clinical Tip:*** It may be useful to think in terms of high-risk families rather than just considering high-risk individuals. Screening children of adults with multiple risk factors might turn up similarly affected children. Having the entire family focus on smoking cessation, control of weight, better diet, and regular exercise has the potential to improve cardiovascular health in dramatic fashion.

Women, Lipids, and CHD

Approximately 50% of all deaths from CHD occur in women. The onset of CHD in women occurs, on average, about 10 years later than in men. As women age, however, the rates of CHD increase; in the postmenopausal woman, the risk from CHD and stroke far exceeds that from cancer. Risk factors in women are similar to those seen in men, although there are some important differences.

Triglycerides are not shown in multivariate analysis to be independent predictors of risk of CHD in men, although they have been shown in some analyses to be independent risk factors in women. HDL-c is a major risk factor in women, and low HDL-c levels are important predictors of CHD mortality in older women. The Lipid Research Clinics Follow-up Study showed that risk of CHD increased especially at levels less than 50 mg/dL. Lp(a) levels in some, but not all, studies are a risk factor for premature CHD. Certainly Lp(a) levels rise after the menopause.

Diabetes is a particularly devastating disease in women, as affected women have the same risk of CHD as their male counterparts. With other risk factors such as high cholesterol and hypertension, there is a 10-year lag between events in affected women as compared with affected men. Diabetic women who smoke are particularly at risk for early sudden death.

Estrogen status must be strongly considered because of the important effects of estrogen not only on the lipid profile, but also on the vessel wall and other risk factors. After menarche, women have higher HDL-c levels than men of similar ages. When women enter the menopause, they often have a rise in weight, higher levels of LDL-c, and lower levels of HDL-c. Oral estrogen replacement therapy is useful in improving the lipid profiles in women with falls in LDL-c and rises in HDL-c. One study showed that oral therapy with 2 mg of micronized estradiol or 0.625 mg of conjugated estrogens lowered LDL-c and raised HDL-c by 14% to 16% (Walsh and Sacks, 1993). The decrease in LDL particles results from accelerated catabolism. Triglycerides rise even more with oral estrogen therapy; this seems to reflect the production of large triglyceride-rich particles of very low-density lipoprotein (VLDL). On the other hand, when modest doses of transdermal estrogen are used, the hepatic exposure to supraphysiologic doses of estrogen is lost and lipids do not change in significant fashion. The introduction of raloxifene, a nonsteroidal benzothiophene, as an estrogenlike drug that does not stimulate the endometrium is a potentially attractive alternative to postmenopausal hormone replacement. Raloxifene in doses that increase bone mineral density lowers total and LDL-c without significant change in triglycerides or HDL-c (Delmas et al, 1997).

> ***Clinical Tip:*** Transdermal estrogen patches are not useful in improving abnormal values for LDL-c and HDL-c in postmenopausal women. On the other hand, for women who have an underlying lipid disorder (and often accompanying obesity), the transdermal patch does not cause the marked increase in triglycerides that is seen with oral estrogen. In women who have significantly elevated triglycerides, addition of oral estrogen or, occasionally, tamoxifen can cause a marked jump in triglycerides to over 1000 mg/dL and increase the risk of pancreatitis (Glueck et al, 1980). Thus a fasting lipid profile should be obtained before women begin postmenopausal estrogen replacement. Those with triglycerides over 300 mg/dL should consider diet, exercise, and weight loss to decrease the triglyceride values before beginning oral estrogen alone (Granfone et al, 1992).

Progestins and estrogen have essentially opposite actions as far as the lipid profile is concerned (although if given alone, progestins can lower HDL-c and LDL-c together). In addition, those progestins with more androgenic activity lower HDL-c more. In the Postmenopausal Estrogen/Progestin Intervention (PEPI) trial, however, micronized progesterone actually improved HDL-c. Finally, there is great concern about the negative effects of progestin on the arterial wall and the oxidation of LDL.

Oral contraceptives have evolved to offer reduced doses of estrogen and progestin. It appears that the relative risk of cardiovascular disease is not increased by past use. Yet, oral contraceptives may affect the lipid profile adversely. Obese women with a familial lipid disorder might develop severely elevated triglyceride levels. Women with multiple risk factors for CHD who take a formulation whose progestin has high androgenic activity can have their HDL-c values lowered. As a general rule, physicians should strongly encourage careful diet and activity to minimize any adverse lipid effects. It would seem prudent to check lipid levels in those with either a family history of premature CHD or obesity. Cigarette smoking should always be discouraged; but for those on oral contraceptives, it is particularly important that the patient understand the marked increase in risk of vascular events that may occur.

A reasonable question to ask is whether aggressive antiatherosclerosis treatment is beneficial for women. To look at this in greater detail, it is useful to divide women into three categories:

- Those who are healthy and at low risk (most premenopausal women)
- Those who have two or more risk factors
- Those with known CHD.

In the first category, careful review of primary prevention trials shows no compelling evidence that treatment of high cholesterol reduces total mortality. Within this essentially low-risk group, however, there may be women with FH. For middle-aged women with FH and one risk factor, aggressive treatment of the hypercholesterolemia is actually cost-effective (Goldman et al, 1993). Moreover, weight gain after 18 years of age leads to increased risk of CHD, so that for women in this group, general advice about a Step

I diet, regular exercise, and smoking cessation would seem to be prudent (Manson et al, 1995).

The above advice is also important for those women with two or more risk factors for CHD but no evidence of overt CHD. This group includes female diabetics and those with insulin-resistance syndromes. The decision to use drug therapy to improve the lipid profile (goals would be LDL-c <130 mg/dL, HDL-c >50 mg/dL, and triglycerides <150 mg/dL) must be individualized, with those at the highest risk being more likely to benefit than those who are at lower levels of risk.

The advisability of treating high-risk women is supported by the results of the Air Force/Texas Coronary Atherosclerosis Prevention Study (AFCAPS/TexCAPS) which showed the beneficial effects of lovastatin in reducing coronary events. Women recruited for this trial had no overt atherosclerotic cardiovascular disease, were 55 years of age of older, had LDL-c >130 to 190 mg/dL, and an HDL-c <47 mg/dL. Yet the low cardiovascular death rates in women (seen in analyses of the Scandinavian Simvastatin Survival Study [4S] and Cholesterol and Recurrent Events [CARE]) suggest caution in considering treatment of women with statins. In this group, results (particularly serial) from noninvasive testing such as electron-beam computed tomography (EBCT) or carotid ultrasonographic imaging might prove useful in determining a group that would benefit from early, aggressive lowering of LDL-c.

In the third category are women with known CHD. Data from trials mentioned before show that FH women who are asymptomatic but have coronary disease revealed on angiography, as well as symptomatic women after a myocardial infarction (MI), as seen in the 4S, the CARE trial, and the Long-Term Intervention with Pravastatin in Ischemic Disease (LIPID) trial, benefit from aggressive lipid-lowering drug therapy. In 4S, women had a significant reduction in cardiovascular end points with simvastastin. In CARE, this reduction was seen as well. In the Canadian Coronary Atherosclerosis Intervention (CCAIT) angiographic trial, women benefited from lovastatin use. Finally, in the Lipoprotein and Coronary Atherosclerosis Study (LCAS), women benefited from fluvastatin. Thus gender should not influence the decision regarding the need for lipid-lowering drug therapy in patients with known CHD.

Clinical trial data now indicate that estrogen/progestin therapy should not be considered ahead of lipid lowering to reduce risk in women with CHD. The Heart and Estrogen/Progestin Replacement Study (HERS) was a blinded, randomized, placebo-controlled clinical trial comparing the effect of daily 0.625 mg of conjugated equine estrogens and 2.5 mg of medroxyprogesterone acetate on the primary outcome of fatal and nonfatal MI (Hulley et al, 1998). Although the rate of new coronary events was similar by the end of the 4.1 years of the study, in the 27,634 postmenopausal women with known coronary artery disease, there was a striking increase in events during the first year in the hormone replacement group. This group also had a significant increase in venous thromboembolic events (Grady et al, 2000).

While explanations for this surprising result are still being sought, an important clue may be the role of Lp(a) (see Chapter 2, *Pathophysiology of Hyperlipoproteinemias*, and Chapter 3, *Classification of Familial Hyperlipidemia*). In this study, Lp(a) was an independent predictor of CHD events and the hormone replacement therapy lowered Lp(a). Moreover, large reductions in Lp(a) were associated with a decreased risk for recurrent CHD. An important interaction was noted; women with initial high Lp(a) had lower rates for CHD if they were randomized to hormone therapy as compared with those on placebo, and those with low initial Lp(a) levels had higher CHD rates in the hormone therapy group compared with the placebo group (Shlipak et al, 2000).

The HERS trial result was surprising because of prior epidemiologic, animal, and small clinical studies suggesting benefit from estrogens. Also, a 3-year study in 875 postmenopausal women from 45 to 64 years of age suggested that estrogen/progestin use had overall a favorable effect on CHD risk factors (PEPI Trial, 1995; see Chapter 4, *Causes of Secondary Hyperlipoproteinemia*). The lesson, however, seems clear. Clinical trial data cannot always be predicted from evidence gleaned from observational studies. This has been true of vitamin E and beta carotene as well.

> ***Clinical Tip:*** It seems prudent to check cholesterol, triglyceride, and HDL-c (and get LDL-c by calculation) in women who will start postmenopausal hormone therapy. If triglycerides are over 300 mg/dL, consideration should be given to nonpharmacologic interventions, such as regular exercise and a weight-reducing diet low in simple sugars. Also, a search for secondary causes and family screening for familial lipid disease would be reasonable. Although only a small group of women who take postmenopausal estrogen develop severe hypertriglyceridemia and/or pancreatitis, the number could be sharply reduced if this precaution were followed.

> ***Clinical Tip:*** Women with breast cancer who are placed on tamoxifen can expect a lowered level of LDL-c but probably no significant change in HDL-c. In women with underlying hypertriglyceridemia, tamoxifen can exacerbate that condition similar to the effect of oral estrogen.

While space limitations preclude a complete coverage of the pros and cons of estrogen therapy for postmenopausal women, it is clear that the risks of estrogen therapy should be carefully considered. For those women who require associated progestin therapy, it should be recognized that this may reduce the benefits that might accrue if estrogen were given alone. Each woman should consider how to respond if cancer of the breast or uterus occurs. Moreover, the risks of these events should be stated and put into perspective. In addition, the patient should be aware that estrogen therapy is associated with changes that make the bile more lithogenic and with a small but finite increase in venous thromboembolism.

> ***Clinical Tip:*** When a uterus is removed, it is considered good clinical practice to stop the associated progestin therapy and just continue with estrogen replacement therapy. In individuals with a familial lipid disorder, this may lead to marked rises in triglycerides in response to oral estrogen, rises which were previously inhibited by the associated progestin therapy (Stone et al, 1994).

More definitive recommendations await the final results of the large Women's Health Initiative (WHI) trial. Nonetheless, in a letter sent to participants in March 2000, the WHI Data and Safety Monitoring Board informed the study participants who were free of CHD at entry of an insignificant trend for more heart attacks, strokes, and venous thromboembolic events in the first 2 years after starting hormone replacement therapy compared with placebo. This finding did not warrant stopping the trial. It would seem to indicate that hormone replacement therapy is not indicated to reduce cardiovascular events in postmenopausal women and that decisions about using postmenopausal hormone replacement should be based on factors other than improvement in CHD risk.

The Elderly, Lipids, and CHD

There is a paucity of clinical trial data to guide the clinician contemplating lipid-lowering therapy in those 70 years and older. Yet cardiovascular disease is the leading cause of death and a major cause of disability in older men and women. Intensive treatment of high-risk elderly patients is an option that must be carefully considered. Several observations suggest that lipid-lowering therapy may prove useful in selected older men and women.

First, intervention in nonlipid risk factors that predict CHD in younger subjects also appears to be effective in the elderly. In the Coronary Artery Surgical Study (CASS), those who quit smoking had lower death rates and a lower incidence of MI than those who continued to smoke. The Swedish Trial in Old Patients With Hypertension (STOP-Hypertension) showed that treatment of elevated systolic blood pressure in elderly men and women 70 to 84 years of age was associated with significant reductions in primary cardiovascular and cerebrovascular end points.

Next, secondary prevention studies of lipid lowering have shown no important decrease in benefit in patients over age 60 as compared with younger patients. The 4S study enrolled patients up to age 70 and the CARE study enrolled patients up to age 75. Thus, by study's end, both trials did have a small group of patients well into the eighth decade of life (see Chapter 7, *Clinical Trials*).

Third, the potential for benefit is greatly increased in the elderly due to what is called "attributable risk." Attributable risk is the excess risk derived by subtracting the risk of those in the lowest percentile for cholesterol from the risk of those in the highest percentile for cholesterol. A closely related statistical attribute called "relative risk" is derived by dividing the risk of those in the highest percentile of cholesterol by the risk of those in the lowest percentile of cholesterol. For example, younger hypercholesterolemic persons have a high relative risk of CHD. Their attributable risk is low because of the low prevalence of CHD in the younger age range. In the elderly, relative risk for hypercholesterolemia is small (presumably because those with the highest cholesterol values have died). On the other hand, attributable or excess risk is high because so many events of CHD occur in the older age groups. One Framingham analysis suggested that relative risk decreases with age and finally inverts at age 80.

> ***Clinical Tip:*** Treatment of high-risk patients in the older age groups presents an opportunity for benefit because this is when the majority of events of CHD occur! This does not mean all older patients with high cholesterol should be treated with medication in a reflex manner. Again, the key emphasis must be on the imperfect, but important, calculation of near-term risk of CHD.

Fourth, elevated cholesterol levels do predict CHD in some studies in the elderly. Overall, it appears that blood cholesterol loses predictive power as people age, and other lipid measurements must be examined more carefully. Recent data emphasize the importance of HDL-c over total cholesterol in predicting mortality from CHD in those over age 70. Castelli noted that high-risk patients in Framingham could be identified by values for HDL-c <40 mg/dL and triglycerides >150 mg/dL. The Cardiovascular Health Study (CHS) measured lipids and lipoproteins in older Americans over age 65 and noted that triglyceride and HDL-c, and to a lesser extent LDL-c, were associated with potentially important modifiable factors, which included obesity, glucose intolerance, renal function, and medication use.

While random screening for high cholesterol in the elderly makes little clinical sense, certain high-risk older people should be considered for active lipid-lowering therapy if they have abnormal lipid profiles. Treatment intensity of high cholesterol requires careful consideration of CHD risk as well as the level of LDL-c and HDL-c. Therapy for those who would derive little benefit (and/or for whom there would be increased risk of side effects or toxicity) should be avoided, and so assessment of comorbidity and functional status is an important part of the clinical decision. Debilitating or disabling illness, malignant disease, or greatly impaired functional status indicates that lipid lowering is unlikely to produce benefit and thus should be avoided.

In patients for whom treatment is considered, secondary causes, such as hypothyroidism, nephrotic syndrome, diabetes, liver disease and kidney disease, should always be considered before starting treatment. Certainly, in the active 70-year-old with hypercholesterolemia who has recently undergone coronary bypass surgery or angioplasty, intensive lipid lowering seems reasonable due to the greater likelihood of earlier recurrence if treatment is withheld solely due to age.

For those with hypercholesterolemia who are over 80 years of age, the decision for lipid-lowering therapy may have clinical merit in high-risk situations, but it should be individualized due to the lack of clinical trial data to guide the decision. Markers of subclinical coronary disease might prove useful, but emerging data suggest limitations. For example, C-reactive protein (CRP) reflects inflammation and predicts CHD risk in middle-aged men and women. In three elderly Finnish cohorts aged 75, 80, and 85 years, CRP was found to be associated with body mass index, insulin, smoking, and inversely, with HDL-c. C-reactive protein predicted overall and cardiovascular mortality only in the 75-year-old cohort (Strandberg and Tilvis, 2000). It is reasonable to continue lipid-lowering therapy in those over 80 when it was begun at a younger age.

> ***Clinical Tip:*** Treating high cholesterol is never done in isolation. Secondary prevention has the highest priority in older patients with known CHD as well as in younger ones. Smoking cessation, hypertension control, increased regular physical exercise, and loss of excess weight are strongly advised.

> ***Clinical Tip:*** A cholesterol-lowering diet is part of the treatment program of high blood cholesterol at any age. Nonetheless, caution is advised in older patients as many of them have limited sources of food choices. Stereotyped low-fat diets can result in deficiencies of protein, calcium, and other nutrients. Dietary counseling in older individuals is often needed to ensure a balanced and nutritious, as well as a cholesterol-lowering, diet.

Obesity

An interest in the lipoprotein disorders of obesity stems from the observation that overweight is associated with increased total mortality and that increased cardiovascular mortality contributes importantly to this problem.

■ Overweight and CHD Risk

The increased risk of mortality associated with excess weight extends over the entire range from very thin to very heavy. This increase in risk is seen over the entire spectrum of ages studied, but it is of greater magnitude among younger individuals (Stevens et al, 1998). In the Nurses' Health Study, this was shown for women; slightly more than 115,000 women were enrolled in this study. The 12-year risk of dying was 2.2 times higher for women with a body mass index (BMI) of 32.0 (equivalent to 175 lbs and 62 inches tall) than for women with a BMI <19.0 (equivalent to 104 lbs and 62 inches tall). The risk gradient for death due to CHD (Figure 11.1) was much steeper; those with a BMI ≥32.0 were at 5.8 times the risk of those with a BMI <19. Similar results are seen in other data involving men.

FIGURE 11.1 — 12-YEAR RELATIVE RISK OF CORONARY HEART DISEASE DEATH: NONSMOKERS IN THE NURSES' HEALTH STUDY

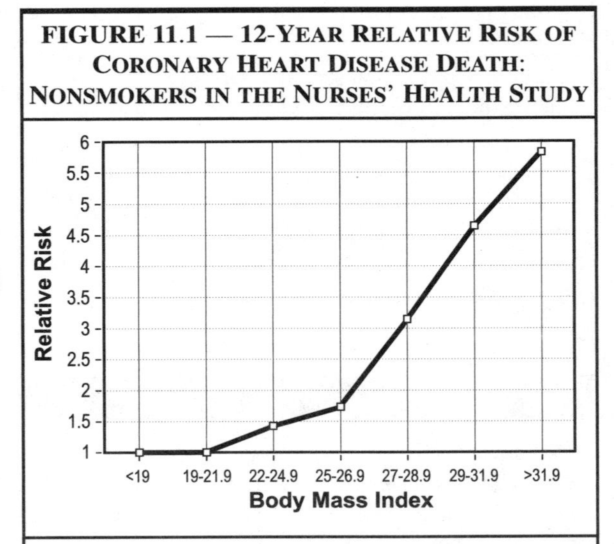

12-year relative risk of death from coronary heart disease among nonsmoking women in the Nurses' Health Study according to body mass index (weight[kg]/height[m]^2).

Data taken from Manson JE, et al. *N Engl J Med.* 1995;333:677-685.

In accord with these data on mortality from CHD, the National Health and Nutrition Evaluation Survey (NHANES) III has demonstrated that obesity is associated with an increased prevalence of CHD in men and in women (Must et al, 1999).

Some previous analyses had identified a U-shaped relationship between mortality and measures of obesity, with a major increase in risk for the thinnest members of the population. However, the conclusion that being thin causes increased risk is erroneous. This erroneous conclusion stems from:

- Failure to control for cigarette smoking (smokers tend to be thinner)
- Inappropriate control for biologic effects of obesity such as hypertension, diabetes, and hyperlipidemia
- Failure to account for weight loss that has been caused by serious disease (Manson et al, 1987).

The increased risk associated with obesity appears to be due not only to changes in lipoprotein metabolism (see below). Weight gain is associated with a clustering of risk factors termed "the metabolic syndrome"; these risk factors include hypercholesterolemia, low HDL cholesterol, glucose intolerance, and hypertension (Wilson et al, 1999; Grundy, 1999).

Overweight appears to confer coronary risk independent of its association with hypertension, hypercholesterolemia, and glucose intolerance, as demonstrated in the 26-year follow-up in the Framingham Study. Additionally, gaining weight after early adulthood appears to contribute independently to CHD risk.

■ Distribution of Body Fat and CHD Risk

A growing body of evidence indicates that the impact of obesity on health is determined to a great extent by the location of excess body fat. In the Nurses' Health Study, for example, waist-to-hip ratio

(ratio of body circumference at waist to that at the hip) was used as an index of visceral fat deposition. This measurement was a better predictor of CHD risk than body mass index. The 12-year risk of developing CHD for those in the highest quintile of waist-to-hip ratio was 8.7 times that for those in the lowest quintile. Most other studies have similarly shown that abdominal obesity is a better predictor of risk.

■ Lipoprotein Disorders Associated With Obesity

Much of the adverse impact of obesity on risk of CHD can be accounted for by the effects of obesity on lipoprotein metabolism. Obesity is associated with undesirable changes in VLDL, LDL, and HDL (Table 11.3).

TABLE 11.3 — EFFECTS OF OBESITY ON PLASMA LIPOPROTEIN LEVELS
• Very low-density lipoprotein (fasting triglyceride) increased
• Low-density lipoprotein (LDL) cholesterol increased
• High-density lipoprotein cholesterol reduced
• Atherogenic pattern of small, dense LDL fostered

Obesity may have a striking impact on serum levels of VLDL and, therefore, on fasting triglycerides. Marked hypertriglyceridemia in an obese individual can often be effectively treated by loss of weight. Furthermore, body weight is a major factor in the expression of type III hyperlipoproteinemia (dysbetalipoproteinemia).

In the Lipid Research Clinics Prevalence Study, BMI was significantly correlated with high triglyceride levels, high LDL levels, and low HDL levels. For 2692 adult males in this study, BMI accounted for 20% to 27% of the population variance in HDL-c after adjustment was made for covariation of BMI with age, cigarette use, exogenous sex hormone use, and alcohol consumption. Among the 2413 adult females, BMI accounted for 17% to 27% of the variation in HDL-c levels after accounting for the impact of the same covariates. The univariate correlation coefficient for BMI with log of triglyceride was approximately 0.3 for both men and women.

Similar results are seen in data from the NHANES II with data obtained 1976 to 1980 (Denke et al, 1993 and 1994). Figure 11.2 demonstrates the relationship with BMI seen for LDL and HDL levels among young adult men in the NHANES II data. The inverse rela-

FIGURE 11.2 — RELATIONSHIP OF BMI TO LDL AND HDL: NHANES II DATA FOR MEN 20 TO 44 YEARS OLD

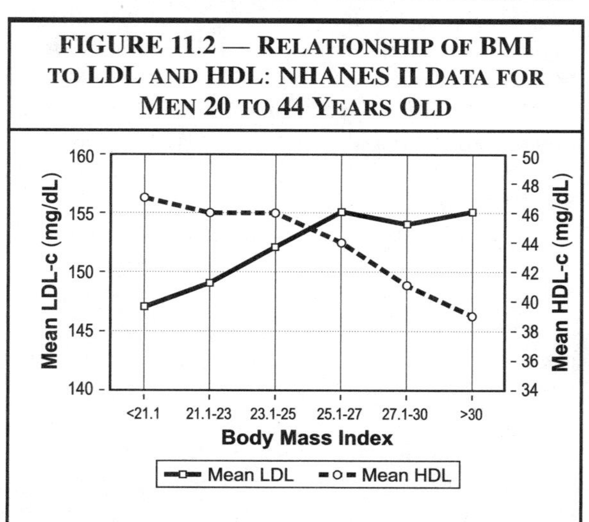

Relationship of body mass index to LDL and HDL cholesterol levels in 644 men aged 20-44 years studied in NHANES II.

Abbreviations: BMI, body mass index; LDL, low-density lipoprotein, HDL, high-density lipoprotein; NHANES, National Health and Nutrition Education Survey.

Data taken from Denke MA, et al. *Arch Intern Med.* 1993;153: 1093-1103.

tionship with HDL was strong in all age groups and in both genders. The relationship with LDL was strongest in young men and in young and middle-aged women. Among overweight young men (BMI >25.0), total serum cholesterol level averaged 19 mg/dL greater than in those with a BMI of 25.0. Stamler calculated that for this group, the elevated cholesterol level associated with overweight would result in a 35% increase in CHD events. He further concluded that "with prevention of overweight and overweight-related higher cholesterol, CHD mortality over the next 15 or more years for the whole cohort of younger men would be lowered by 14%." Recent analyses have shown a frightening 1.5-fold increase in the prevalence of obesity during the past decade (Mokdad et al, 1999). Thus the benefits resulting from prevention of obesity should be considerably greater than the 14% reduction in CHD mortality calculated by Stamler using older data.

The adverse impact of obesity on LDL-c and HDL-c levels was confirmed in the Framingham Offspring Study. These findings were extended with the observation that increasing BMI led to an increased prevalence of the atherogenic LDL pattern B (predominance of small, dense LDL) (Figure 11.3).

The mechanisms producing the lipoprotein abnormalities seen with obesity are complex. Obesity leads to insulin resistance. The impact of insulin resistance on lipoprotein physiology is discussed in detail in the following section of this chapter. Insulin resistance leads to increased hepatic secretion of VLDL, and this is manifested by elevated fasting triglyceride levels. Furthermore, obesity is associated with reduced lipoprotein lipase activity and, thereby, with reduced clearance of VLDL from plasma.

Since LDL is a metabolic product of VLDL, one might expect obesity to be associated with some elevation of LDL levels. Additionally, the hypertriglyc-

FIGURE 11.3 — FRAMINGHAM OFFSPRING STUDY: BMI VS LDL PATTERN B IN NONSMOKERS

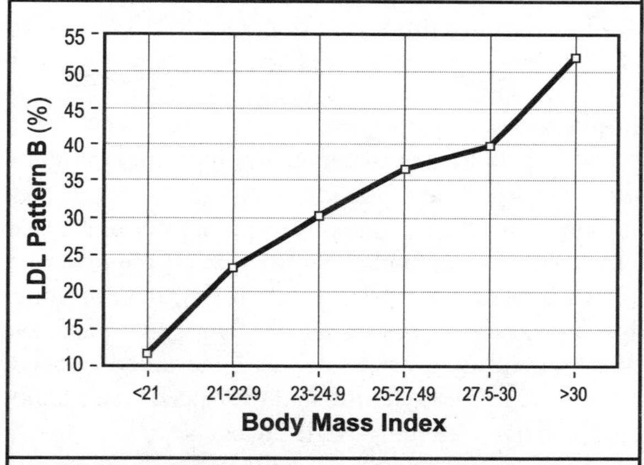

Relationship of body mass index to prevalence of LDL pattern B (small dense LDL) in the Framingham Offspring Study.

Abbreviations: BMI, body mass index; LDL, low-density lipoprotein.

Data taken from Lamon-Fava S, et al. *Arterioscler Thromb Vasc Biol*. 1996;16:1509-1515.

eridemia occurring in obesity will reduce HDL levels by mechanisms involving cholesteryl ester transfer protein (see Chapter 2, *Pathophysiology of Hyperlipoproteinemias*).

Excess body weight attenuates the impact of dietary cholesterol (but not of saturated fats) on serum cholesterol levels (Goff et al, 1993). Thus reduction of dietary cholesterol will lead to greater reduction in serum cholesterol of overweight individuals after reduction of weight.

A number of observations suggest the particular importance of visceral adipose tissue in influencing lipoprotein physiology (Table 11.4). Visceral adipocytes seem to be especially sensitive to stimulation of lipolysis and to have reduced sensitivity to the antilipolytic effects of insulin. Additionally, lipolysis in visceral adipocytes deposits free fatty acids (FFAs) directly into the portal blood stream and, therefore, increases hepatic exposure to FFAs. This, in turn, increases the hepatic secretion of VLDL. The foregoing indicates mechanisms by which visceral obesity may increase the secretion of triglyceride-rich lipoproteins. Additionally, visceral obesity retards the clearance of postprandial triglyceride-rich lipoproteins (Taira et al, 1999; Mekki et al, 1999), thus prolonging the residence in plasma of atherogenic particles. Visceral obesity is associated with increased levels of hepatic triglyceride lipase, and this speeds the catabolism of HDL, reducing HDL levels.

TABLE 11.4 — IMPACT OF VISCERAL OBESITY ON LIPOPROTEIN PHYSIOLOGY
• Increased sensitivity of visceral adipocytes to lipolytic stimuli
• Reduced sensitivity of visceral adipocytes to antilipolytic effects of insulin
• Increased hepatic exposure to free fatty acids with lipolysis in visceral location
• Increased hepatic triglyceride lipase activity

Insulin Resistance, Dyslipidemia, and Coronary Heart Disease

Hyperinsulinemia and insulin resistance have been reported to be associated with CHD and with other metabolic disorders that themselves predispose to development of CHD.

■ Association of Insulin Resistance With Coronary Risk

The reports of association of insulin resistance with CHD have not been consistent (Wingard et al, 1995). In virtually all of these studies, measurement of elevated insulin levels was used as a surrogate for insulin resistance. Most of the studies which reported an absence of association between hyperinsulinism and CHD were hampered by a small number of CHD events; this limited their power to detect a relationship. Two recent reports of large studies do indicate that hyperinsulinism is associated with increased CHD risk and that this risk is independent of associations with established CHD risk factors (Despres et al, 1996; Perry et al, 1996). However, a nested case control study from the very large population of the Multiple Risk Factor Intervention Trial (MRFIT) did not show any relationship between fasting serum insulin levels and CHD risk (Orchard et al, 1994). The question of whether hyperinsulinism (fasting or after a glucose load) is an independent risk factor for CHD thus remains controversial.

■ Association of Hyperinsulinism With Coronary Risk Factors

In addition to any independent effect on CHD risk, it is clear that hyperinsulinism is associated with other abnormalities which themselves increase risk of CHD (Table 11.5). Hyperinsulinism is associated with increased prevalence of established risk factors, including reduced HDL-c, hypertension, and obesity. Insulin resistance and hyperinsulinism are felt to be central to the "metabolic syndrome," in which there is a clustering of several risk factors ([a]Grundy, 1999). In addition to these very well-established independent risk factors for CHD, hyperinsulinism is also associated with a predominance of small, dense LDL par-

TABLE 11.5 — CORONARY HEART DISEASE RISK FACTORS ASSOCIATED WITH HYPERINSULINISM
• Low high-density lipoprotein cholesterol concentrations • High low-density lipoprotein (LDL) cholesterol concentrations • Small, dense LDL particles • Hypertension • Hypertriglyceridemia • Abdominal obesity

ticles, elevated triglyceride levels, and impaired hemostasis. It is not clear whether insulin resistance has an impact on LDL-c levels; reports in the literature have not been consistent.

■ Mechanisms by Which Insulin Resistance Influences Lipoproteins

Insulin resistance can influence lipoprotein metabolism in a variety of ways (Table 11.6).

In a transgenic mouse model, insulin resistance in skeletal muscle resulted in increased body fat and increased plasma levels of free fatty acids and of triglycerides (Moller et al, 1996). Thus insulin resistance per se can influence lipoprotein physiology.

Insulin resistance affects the metabolism of triglyceride-rich lipoproteins (VLDL and chylomicrons). Crucial events are effects on hormone-sensitive lipase (HSL), the enzyme responsible for hydrolysis of triglyceride stores in adipose tissue, and LPL, the enzyme responsible for hydrolysis of triglyceride in circulating VLDL and chylomicrons. HSL is suppressed by insulin. Thus, in insulin-resistant states, lipolysis in adipose tissue is stimulated, and there is an increased flux of FFA to the liver. This provides substrate for synthesis of triglyceride and leads to the se-

TABLE 11.6 — INFLUENCES OF INSULIN RESISTANCE ON LIPOPROTEIN METABOLISM

Primary Effect	Impact on Lipoproteins
VLDL Metabolism	
↑ Flux of FFA to liver	↑ Secretion of TG and increased VLDL apo B-100 secretion
↓ Adipose tissue LPL activity	↓ Clearance of VLDL from circulation
LDL Metabolism	
↑ Secretion of VLDL apo B-100	↑ LDL synthesis (because of increased precursor)
↓ Adipose tissue LPL activity	↓ Conversion of VLDL to LDL
↓ LDL receptor activity	↓ Clearance of LDL from plasma
HDL Metabolism	
Hypertriglyceridemia	↑ HDL catabolism by mechanisms involving CETP
↑ Hepatic TGL activity	↑ Catabolism of HDL, TG, and PPL
↓ LPL activity	↓ HDL synthesis due to reduced production of "surface remnants" of VLDL and CM

Abbreviations: VLDL, very low-density lipoprotein; FFA, free fatty acid; TG, triglyceride; apo, apolipoprotein; LPL, lipoprotein lipase; LDL, low-density lipoprotein; HDL, high-density lipoprotein; CETP, cholesteryl ester transfer protein; TGL, triglyceride lipase; PPL, phospholipid; CM, chylomicron.

cretion of increased amounts of VLDL triglyceride and apo B-100 by the liver.

Furthermore, LPL activity is stimulated by insulin; therefore, in insulin-resistant states, adipose tissue activity of LPL is reduced, and the clearance of VLDL and chylomicrons from the circulation is retarded.

Some metabolic effects of insulin resistance increase LDL levels, while other effects of insulin resistance reduce LDL levels:

- Enhanced secretion of apo B-100 in VLDL provides increased precursor for the production of LDL.
- Reduced LPL activity slows the conversion of VLDL to LDL and may result in reduced production of LDL.
- Reduced activity of LDL receptors leads to retarded clearance of LDL from plasma and increases LDL levels (insulin stimulates LDL receptor activity).

Insulin resistance can reduce HDL levels by several mechanisms. Increased plasma levels of triglyceride-rich lipoproteins (generated as indicated above) will reduce HDL levels by mechanisms involving the cholesteryl ester transfer protein (see Chapter 2, *Pathophysiology of Hyperlipoproteinemias*). Additionally, insulin-resistant states are associated with elevated activity of hepatic triglyceride lipase, and this may speed the catabolism of HDL in plasma. The reduced activity of adipose tissue LPL seen with insulin resistance may lead to reduced synthesis of HDL. As noted in Chapter 2, a portion of the synthesis of HDL is attributed to the production of "surface remnants" of triglyceride-rich lipoproteins by the action of LPL.

Diabetes Mellitus, Dyslipidemia, and Coronary Heart Disease

For an extensive review of issues relating to the management of lipid disorders in diabetes, see Haffner et al (1998) and [a]Grundy et al (1999).

■ Diabetes and Coronary Risk

Diabetes carries with it a great burden of atherosclerotic risk. In 1974, the 16-year follow-up experience of the Framingham Study was reported; the risk of CHD death among diabetics was 3.2 times that of nondiabetics. The increased risk persists after adjustment is made for other known risk factors, and the impact of diabetes on relative risk of CHD is greater for women than for men. Diabetes eliminates the female advantage in CHD rates. The increased relative risk imposed by diabetes in women appears to be due at least in part to increased risk in women with low HDL-c (≤50 mg/dL) and/or high VLDL-c (≥20 mg/dL) levels (Goldschmid et al, 1994).

Among patients with type 1 diabetes mellitus, the Joslin Clinic experience showed a cumulative CHD mortality of 35% by age 55 years (Krolewski et al, 1987). Among the living patients with type 1 diabetes aged 45 to 59 years, the prevalence of a positive electrocardiographic stress test was 33%. These data show that a 45-year-old patient with type 1 diabetes has a 39% chance of dying of CHD over the coming 15 years (Figure 11.4).

Type 2 diabetes also exacts a severe toll of morbidity and mortality due to CHD. In the Nurses' Health Study (Manson et al, 1991), of 116,117 women who were followed prospectively for 8 years, the risk of nonfatal MI and CHD death among diabetics was 6.7 times that for nondiabetics. Risk of nonfatal MI for those treated with oral hypoglycemic agents (relative risk [RR] 8.7) was similar to that for those treated

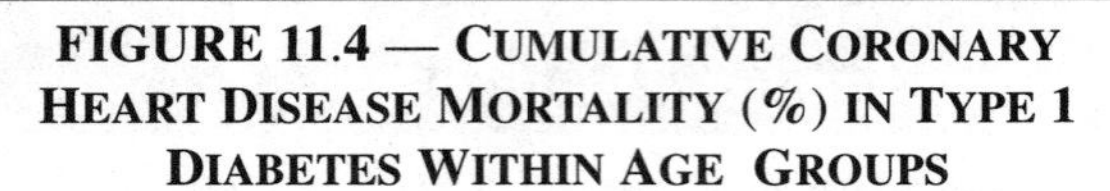

FIGURE 11.4 — CUMULATIVE CORONARY HEART DISEASE MORTALITY (%) IN TYPE 1 DIABETES WITHIN AGE GROUPS

Cumulative coronary heart disease (CHD) mortality within age groups of type 1 diabetes patients.

Data calculated from Krolewski AS, et al. *Am J Cardiol.* 1987; 59:750-755.

with insulin (RR 8.2); it was somewhat lower for those treated with diet alone (RR 5.7), probably reflecting reduced severity of disease.

Coronary heart disease mortality in type 2 diabetics without a prior history of MI is as high as that in nondiabetics who have had prior MI ([a]Haffner, 1998). This provides some of the rationale for particularly energetic management of risk factors in diabetics.

Even mild, asymptomatic hyperglycemia leads to increased coronary risk. When serum glucose repeatedly exceeds 200 mg/dL in individuals without previously known diabetes within 1 hour after a 50-g oral glucose challenge, the 19-year risk of CHD death increases 2.4-fold compared to those with lower serum glucose levels (Vaccaro et al, 1992).

■ Lipoprotein Disorders in Diabetes

Atherogenic disorders of plasma lipoproteins have been associated with asymptomatic hyperglycemia, diagnosed type 1 diabetes, and type 2 diabetes.

The major lipoprotein abnormalities of type 2 diabetes are low HDL concentrations, high triglyceride concentrations, and prevalence of small, dense LDL (Table 11.7). These abnormalities and their pathophysiology are very similar to what is seen in nondiabetics with insulin-resistant states.

TABLE 11.7 — LIPOPROTEIN ABNORMALITIES CHARACTERISTIC OF TYPE 2 DIABETES
• Hypertriglyceridemia (high very low-density lipoprotein) • Low high-density lipoprotein cholesterol concentration • Small, dense low-density lipoprotein (LDL) • Glycosylated LDL

Additionally, glycosylation of lipoproteins in diabetics may enhance the atherogenicity of lipoproteins. Data from NHANES II clearly show a tendency to lower HDL cholesterol levels and higher fasting triglyceride levels in patients with type 2 diabetes (Figure 11.5). This is seen in both men and women, and in blacks and whites. Type 2 diabetes is strongly associated with a predominance of atherogenic small, dense LDL particles (LDL subclass phenotype B). This association is at least partially a result of hypertriglyceridemia in type 2 diabetes; in men, but not in women, the association of type 2 diabetes with small, dense LDL could be entirely accounted for on the basis of hypertriglyceridemia (Austin and Edwards, 1996).

A different pattern is apparent for type 1 diabetes in the recent data from the Diabetes Control and Complications Trial (DCCT). In an analysis involv-

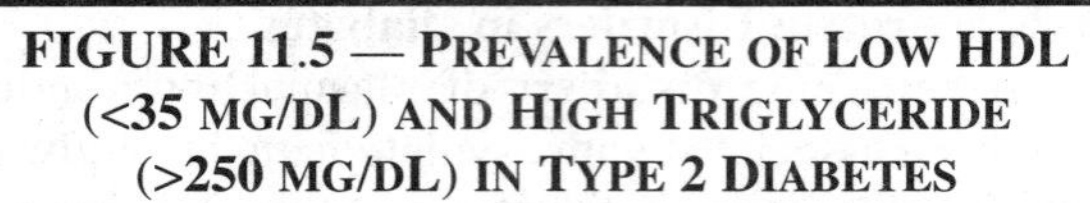

FIGURE 11.5 — PREVALENCE OF LOW HDL (<35 MG/DL) AND HIGH TRIGLYCERIDE (>250 MG/DL) IN TYPE 2 DIABETES

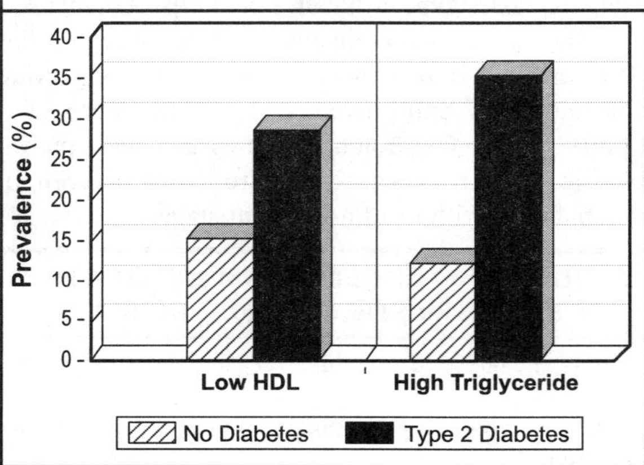

Prevalence of low high-density lipoprotein (HDL) (<35 mg/dL) and high triglyceride (>250 mg/dL) levels in type 2 diabetes.

Data are taken from Cowie CC, et al. *Circulation*. 1994;90:1185-1193.

ing over 1500 patients, the lipid and lipoprotein levels in people with type 1 diabetes were found to be similar to those in the general population (DCCT Research Group, 1992). Previous studies, however, have shown increased LDL levels, low HDL levels, and high triglyceride levels in association with poorly controlled type 1 diabetes (Howard, 1987). The better lipoprotein profiles seen in the more current DCCT data may be caused by:

- A secular trend toward improved glycemic control
- Altered dietary recommendations for diabetics with a current emphasis on restriction of cholesterol and saturated fats.

In untreated type 1 diabetes, a deficiency of LPL causes reduced clearance of VLDL and chylomicrons; thus plasma levels of these triglyceride-rich lipoproteins are markedly increased. FFA levels in plasma are high because of increased activity of hormone-sensitive lipase, the enzyme responsible for hydrolysis of stores of triglyceride in adipose tissue; this enzyme is suppressed by insulin. When insulin is present in the small amount sufficient to allow hepatic lipogenesis and protein synthesis, the increased FFA flux to the liver provides substrate for production of VLDL, and VLDL secretion is increased.

Hyperglycemia, whether due to type 1 or type 2 diabetes, leads to glycosylation of LDL, and this may be atherogenic (Table 11.8) (Lopes-Virella and Virella, 1992). Recognition of glycosylated LDL by the LDL receptor is impaired. Thus glycosylated LDL is cleared more slowly from plasma than is native LDL, and clearance by atherogenic, non-LDL receptor pathways is greater for glycosylated LDL. Macrophages internalize glycosylated LDL at an enhanced rate; thus the glycosylation of LDL fosters the formation of foam cells. Glycosylated LDL is immunogenic, and diabetics have been shown to have circulating complexes of immunoglobulins with glycosylated LDL. These immune complexes are internalized by macrophages in unregulated fashion favoring the formation of foam cells. Additionally, glycosylated LDL has been shown to foster platelet aggregation.

TABLE 11.8 — GLYCOSYLATION OF LDL: POTENTIAL ATHEROGENIC EFFECTS
• Impaired recognition by low-density lipoprotein (LDL) receptor
• Enhanced uptake by macrophages
• Stimulation of platelet aggregation
• Glycosylated LDL is immunogenic

■ Treatment of Lipid Disorders in Diabetes

In general, lipoprotein disorders in diabetics are treated with the same considerations as in nondiabetics. However, certain considerations are specific for diabetics.

Control of glycemia is important in management of diabetic dyslipidemia. Sosenko et al (1980) demonstrated in children with type 1 diabetes that poor control of glycemia (measured as elevated hemoglobin A_{1C}) was associated with increased LDL and VLDL levels. In the DCCT, intensive treatment of type 1 diabetes reduced LDL and fasting triglyceride levels (DCCT Research Group, 1995). Thus in type 1 diabetes, intensive management of diabetes can contribute to control of lipoprotein abnormalities. Similarly, improved control of type 2 diabetes has been shown to lead to improvement in associated lipoprotein disorders. With improved control, VLDL synthesis falls and VLDL clearance is enhanced; the result is a reduction in fasting plasma triglyceride levels.

Since weight loss in the obese and exercise can favorably influence insulin resistance and diabetic control in type 2 diabetes, efforts at weight control and exercise are crucial to treatment of dyslipidemia in type 2 diabetes.

Additionally, and of most importance, the metabolic control of diabetes predicts the frequency of subsequent CHD events. This was seen in the DCCT with type 1 diabetes (DCCT Research Group, 1995) and it is also true for type 2 diabetes (Kuusisto et al, 1994).

Guidelines for Dietary and Pharmacologic Treatment of Lipid Disorders in Diabetic Patients

The management of lipid disorders in diabetics has recently been considered in detail in a scientific statement of the American Heart Association ([a]Grundy, 1999), in a technical review (Haffner, 1998), and in a position statement of the American Diabetes

Association (ADA) (American Diabetes Association, 1998). As in the NCEP guidelines, *the first priority is on reducing elevated levels of LDL-c.* In the 4S (Pyorala et al, 1997) and CARE studies (Grundy et al, 1998), LDL reduction with statins produced benefits (reduced CHD rates, and [in the 4S] reduced mortality) in diabetic patients that were similar to those seen in nondiabetics.

The guidelines of the ADA call for somewhat more intensive LDL-c–reducing treatment than do those of the NCEP; the rationale for somewhat more intensive treatment involves:

- The high incidence of CHD in diabetes
- Particularly high case-fatality rates of diabetics developing CHD.

The guidelines for treatment recommended by the ADA are summarized in Table 11.9.

TABLE 11.9 — GUILDELINES FOR TREATMENT TO REDUCE LDL-C IN DIABETIC PATIENTS

	With Preexisting CHD, PVD, or CVD	Without Preexisting CHD, PVD, or CVD
Dietary Therapy		
Initiation level	>100 mg/dL	>100 mg/dL
LDL goal	≤100 mg/dL	≤100 mg/dL
Drug Therapy		
Initiation level	>100 mg/dL	≥130 mg/dL
LDL goal	≤100 mg/dL	<130 mg/dL

Abbreviations: LDL-c, low-density lipoprotein cholesterol; CHD, coronary heart disease; PVD, peripheral vascular disease; CVD, cerebrovascular disease; LDL, low-density lipoprotein.

For those diabetics with preexisting vascular disease, the guidelines are similar to those of the NCEP. The treatment goal is to reduce LDL-c to 100 mg/dL or less. Drug therapy is considered when LDL-c remains over 100 mg/dL despite intensive dietary efforts. (This approach is somewhat more aggressive than that proposed by the NCEP.)

For those diabetics without preexisting CHD, the ADA recommends that drug treatment be initiated whenever LDL-c levels remain at least 130 mg/dL despite dietary measures and that the goal of this treatment be to reduce LDL-c to levels below 130 mg/dL. (This approach is midway between the NCEP recommendations for patients with vascular disease and for high-risk patients without preexisting vascular disease.) The ADA recognizes that "some authorities recommend an LDL goal ≤100 mg/dL" for diabetic patients with one or more CHD risk factors beyond diabetes.

Because low HDL levels and hypertriglyceridemia appear to be risk factors for CHD in diabetics, the ADA also recognizes *low HDL levels and elevated triglyceride levels as targets of treatment.* Here the approach is first with improved glycemic control, weight loss (for those who are overweight), and exercise. If this is not successful, pharmacologic therapy can then be considered.

Pharmacologic treatment of dyslipidemia in diabetics involves the same medications as are used in nondiabetics. However, there are certain special considerations for the diabetic population.

Since nicotinic acid can increase plasma glucose levels and lead to impaired diabetic control, it should not usually be a first-line drug for treatment of diabetics (Garg and Grundy, 1990).

Bile acid sequestrants (cholestyramine and colestipol) can be as effective in reducing LDL levels of diabetics as in nondiabetics. However, patients with type 2 diabetes often tend to be hypertriglyceridemic and cholestyramine increases plasma triglyceride levels. Thus hypertriglyceridemia may limit the use of this drug in diabetics.

Fibric acid derivatives can increase LDL-c levels in hypertriglyceridemic patients, and diabetic patients tend to be hypertriglyceridemic. Thus this complication of treatment may limit the use of fibric acid derivatives in diabetics. Nonetheless, in the VA HDL Intervention Trial (Rubins, 1999), in which enrollment required HDL ≤40 mg/dL, LDL ≤140 mg/dL, and triglyceride ≤300 mg/dL, treatment with gemfibrozil reduced CHD rates just as effectively in diabetics as in others.

The HMG-CoA reductase inhibitors can be used in diabetics in the same manner as in nondiabetics. Although the major use of these drugs is in reducing levels of LDL-c, in high doses they are also moderately effective in reducing triglyceride levels. As noted above, the HMG-CoA reductase inhibitors have been effective in reducing the frequency of clinical coronary events in diabetic patients in two large clinical trials (CARE and 4S).

SUGGESTED READINGS

Children

Bao W, Srinivasan SR, Wattigney WA, Berenson GS. Usefulness of childhood low-density lipoprotein cholesterol level in predicting adult dyslipidemia and other cardiovascular risks. The Bogalusa Heart Study. *Arch Intern Med.* 1996;156:1315-1320.

Bao W, Srinivasan SR, Wattigney WA, Berenson GS. Persistence of multiple cardiovascular risk clustering related to syndrome X from childhood to young adulthood. The Bogalusa Heart Study. *Arch Intern Med.* 1994;154:1842-1847.

Luepker RV, Perry CL, McKinlay SM, et al. Outcomes of a field trial to improve children's dietary patterns and physical activity. The Child and Adolescent Trial for Cardiovascular Health. CATCH collaborative group. *JAMA.* 1996;275:768-776.

Niinikoski H, Viikari J, Ronnemaa T, et al. Prospective randomized trial of low-saturated fat, low-cholesterol diet during the first 3 years of life. The STRIP baby project. *Circulation.* 1996;94:1386-1393.

Report of the Expert Panel on Blood Cholesterol Levels in Children and Adolescents. National Cholesterol Education Program. US Department of Health and Human Services. NIH Publication No. 91-2732. September, 1991.

Rifkind BM, Schucker B, Gordon DJ. When should patients with heterozygous familial hypercholesterolemia be treated? *JAMA.* 1999;281:180-181. Editorial.

Stein EA, Illingworth DR, Kwiterovich PO Jr, et al. Efficacy and safety of lovastatin in adolescent males with heterozygous familial hypercholesterolemia: a randomized controlled trial. *JAMA.* 1999;281:137-144.

The Writing Group for the DISC Collaborative Research Group. Efficacy and safety of lowering dietary intake of fat and cholesterol in children with elevated low-density lipoprotein cholesterol. The Dietary Intervention Study in Children (DISC). *JAMA.* 1995;273:1429-1435.

Diabetes

American Diabetes Association. Management of dyslipidemia in adults with diabetes. *Diabetes Care.* 1998;21:179-182.

Austin MA, Edwards KL. Small, dense low-density lipoproteins, the insulin resistance syndrome and noninsulin-dependent diabetes. *Curr Opin Lipidol.* 1996;7:167-171.

Cowie CC, Howard BV, Harris MI. Serum lipoproteins in African Americans and whites with non-insulin-dependent diabetes in the US population. *Circulation.* 1994;90:1185-1193.

Garg A, Grundy SM. Nicotinic acid as therapy for dyslipidemia in non-insulin-dependent diabetes mellitus. *JAMA.* 1990;264:723-726.

Goldschmid MG, Barrett-Connor EL, Edelstein SL, Wingard DL, Cohn BA, Herman WH. Dyslipidemia and ischemic heart disease mortality among men and women with diabetes. *Circulation.* 1994;89:991-997.

Haffner SM. Management of dyslipidemia in adults with diabetes. *Diabetes Care*. 1998;21:160-178.

Howard BV. Lipoprotein metabolism in diabetes mellitus. *J Lipid Res*. 1987;28:613-628.

Krolewski AS, Kosinski EJ, Warram JH, et al. Magnitude and determinants of coronary artery disease in juvenile-onset, insulin-dependent diabetes mellitus. *Am J Cardiol*. 1987;59:750-755.

Kuusisto J, Mykkanen L, Pyorala K, Laakso M. NIDDM and its metabolic control predict coronary heart disease in elderly subjects. *Diabetes*. 1994;43:960-967.

Lopes-Virella MF, Virella G. Immune mechanisms of atherosclerosis in diabetes mellitus. *Diabetes*. 1992;41(suppl 2):86-91.

Manson JE, Colditz GA, Stampfer MJ, et al. A prospective study of maturity-onset diabetes mellitus and risk of coronary heart disease and stroke in women. *Arch Intern Med*. 1991;151:1141-1147.

Pyorala K, Pedersen TR, Kjekshus J, Faergeman O, Olsson AG, Thorgeirsson G. Cholesterol lowering with simvastatin improves prognosis of diabetic patients with coronary heart disease. A subgroup analysis of the Scandinavian Simvastatin Survival Study (4S) [published erratum appears in *Diabetes Care*. 1997;20:1048]. *Diabetes Care*. 1997;20:614-620.

Rubins HB, Robins SJ, Collins, D, et al. Gemfibrozil for the secondary prevention of coronary heart disease in men with low levels of high-density lipoprotein cholesterol. Veterans Affairs High-Density Lipoprotein Cholesterol Intervention Trial Study Group. *N Engl J Med*. 1999;341:410-418.

Sosenko JM, Breslow JL, Miettinen OS, Gabbay KH. Hyperglycemia and plasma lipid levels: a prospective study of young insulin-dependent diabetic patients. *N Engl J Med*. 1980;302:650-654.

The DCCT Research Group. Lipid and lipoprotein levels in patients with IDDM. *Diabetes Care*. 1992;15:886-894.

The Diabetes Control and Complications Trial (DCCT) Research Group. Effect of intensive diabetes management on macrovascular events and risk factors in the diabetes control and complications trial. *Am J Cardiol.* 1995;75:894-903.

Vaccaro O, Ruth KJ, Stamler J. Relationship of postload plasma glucose measurements to mortality with 19-year follow-up. Comparison of one versus two plasma glucose measurements in the Chicago Peoples Gas Company Study. *Diabetes Care.* 1992; 15:1328-1334.

Elderly

Benfante R, Reed D. Is elevated serum cholesterol level a risk factor for coronary heart disease in the elderly? *JAMA.* 1990; 263:393-396.

Castelli WP, Wilson PW, Levy D, Anderson K. Cardiovascular risk factors in the elderly. *Am J Cardiol.* 1989;63:12H-19H.

Corti MC, Guralnik JM, Salive ME, et al. HDL cholesterol predicts coronary heart disease mortality in older persons. *JAMA.* 1995;274:539-544.

Dahlöf B, Lindholm LH, Hansson L, Schersten B, Ekbom T, Wester PO. Morbidity and mortality in the Swedish Trial in Old Patients With Hypertension (STOP-Hypertension). *Lancet.* 1991;338:1281-1285 .

Denke M. Drug treatment of hyperlipidemia in elderly patients. *Curr Opin Lipidol.* 1993;4:56-62.

Ettinger WH, Wahl PW, Kuller LH, et al. Lipoprotein lipids in older people. Results from the Cardiovascular Health Study. The CHS Collaborative Research Group. *Circulation.* 1992;86:858-869.

Hermanson B, Omenn GS, Kronmal RA, Gersh BJ. Beneficial six-year outcome of smoking cessation in older men and women with coronary artery disease. Results from the CASS Registry. *N Engl J Med.* 1988;319:1365-1369.

Hulley SB, Newman TB. Cholesterol in the elderly. Is it important? *JAMA.* 1994;272:1372-1374. Editorial.

Rubin SM, Sidney S, Black DM, Browner WS, Hulley SB, Cummings SR. High blood cholesterol in elderly men and the excess risk for coronary heart disease. *Ann Intern Med.* 1990; 113:916-920.

Schaefer EJ, Lichtenstein AH, Lamon-Fava S, et al. Efficacy of a National Cholesterol Education Program Step 2 diet in normolipidemic and hypercholesterolemic middle-aged and elderly men and women. *Arterioscler Thromb Vasc Biol.* 1995;15: 1079-1085.

Stone NJ. The 75-year old patient with hypercholesterolemia: to treat or not to treat? *Nutr Rev.* 1994;52:S31-S33.

Strandberg TE, Tilvis RC. C-reactive protein, cardiovascular risk factors, and mortality in a prospective study in the elderly. *Arterioscler Thromb Vasc Biol.* 2000;20:1057-1060.

Insulin Resistance

Despres JP, Lamarche B, Mauriege P, et al. Hyperinsulinemia as an independent risk factor for ischemic heart disease. *N Engl J Med.* 1996;334:952-957.

Garg A. Insulin resistance in the pathogenesis of dyslipidemia. *Diabetes Care.* 1996;19:387-389.

Goldberg RB, Mellies MJ, Sacks FM, et al. Cardiovascular events and their reduction with pravastatin in diabetic and glucose-intolerant myocardial infarction survivors with average cholesterol levels: subgroup analyses in the cholesterol and recurrent events (CARE) trial. The Care Investigators. *Circulation.* 1998;98:2513-2519.

[a]Grundy SM, Benjamin IJ, Burke GL, et al. Diabetes and cardiovascular disease: a statement for healthcare professionals from the American Heart Association. *Circulation.* 1999; 100:1134-1146.

[a]Haffner SM, Lehto S, Ronnemaa T, Pyorala K, Laakso M. Mortality from coronary heart disease in subjects with type 2 diabetes and in nondiabetic subjects with and without prior myocardial infarction. *N Engl J Med.* 1998;339:229-234.

Moller DE, Chang PY, Yaspelkis BB 3rd, Flier JS, Wallberg-Henriksson H, Ivy JL. Transgenic mice with muscle-specific insulin resistance develop increased adiposity, impaired glucose tolerance, and dyslipidemia. *Endocrinology.* 1996;137:2397-2405.

Orchard TJ, Eichner J, Kuller LH, Becker DJ, McCallum LM, Grandits GA. Insulin as a predictor of coronary heart disease: interaction with apolipoprotein E phenotype. A report from the Multiple Risk Factor Intervention Trial. *Ann Epidemiol.* 1994;4:40-45.

Perry IJ, Wannamethee SG, Whincup PH, Shaper AG, Walker MK, Alberti KG. Serum insulin and incident coronary heart disease in middle-aged British men. *Am J Epidemiol.* 1996; 144:224-234.

Wingard DL, Barrett-Connor EL, Ferrara A. Is insulin really a heart disease risk factor? *Diabetes Care.* 1995;18:1299-1304.

Obesity

Denke MA, Sempos CT, Grundy SM. Excess body weight. An under-recognized contributor to dyslipidemia in white American women. *Arch Intern Med.* 1994;154:401-410.

Denke MA, Sempos CT, Grundy SM. Excess body weight. An underrecognized contributor to high blood cholesterol levels in white American men. *Arch Intern Med.* 1993;153:1093-1103.

Despres JP, Moorjani S, Lupien PJ, Tremblay A, Nadeau A, Bouchard C. Regional distribution of body fat, plasma lipoproteins, and cardiovascular risk. *Arteriosclerosis.* 1990;10:497-511.

Folsom AR, Jacobs DR Jr, Wagenknecht LE, et al. Increase in fasting insulin and glucose over seven years with increasing weight and inactivity of young adults. The CARDIA Study. Coronary Artery Risk Development in Young Adults. *Am J Epidemiol.* 1996;144:235-246.

Glueck CJ, Taylor HL, Jacobs D, Morrison JA, Beaglehole R, Williams OD. Plasma high-density lipoprotein cholesterol: association with measurements of body mass. The Lipid Research Clinics Program Prevalence Study. *Circulation.* 1980;62(suppl 4):IV62-IV69.

Goff DC Jr, Skekelle RB, Moye LA, Katan MB, Gotto AM Jr, Stamler J. Does body fatness modify the effect of dietary cholesterol on serum cholesterol? Results from the Chicago Western Electric Study. *Am J Epidemiol.* 1993;137:171-177.

Grundy SM. Hypertriglyceridemia, insulin resistance, and the metabolic syndrome. *Am J Cardiol.* 1999;83:25F-29F.

Hubert HB, Feinleib M, McNamara PM, Castelli WP. Obesity as an independent risk factor for cardiovascular disease: a 26-year follow-up of participants in the Framingham Heart Study. *Circulation*. 1983;67:968-977.

Lamon-Fava S, Wilson PW, Schaefer EJ. Impact of body mass index on coronary heart disease risk factors in men and women. The Framingham Offspring Study. *Arterioscler Thromb Vasc Biol*. 1996;16:1509-1515.

Manson JE, Stampfer MJ, Hennekens CH, Willett WC. Body weight and longevity. A reassessment. *JAMA*. 1987;257:353-358.

Manson JE, Willett WC, Stampfer MJ, et al. Body weight and mortality among women. *N Engl J Med*. 1995;333:677-685.

Mekki N, Christofilis MA, Charbonnier M, et al. Influence of obesity and body fat distribution on postprandial lipemia and triglyceride-rich lipoproteins in adult women. *J Clin Endocrinol Metab*. 1999;84:184-191.

Mokdad AH, Serdula MK, Dietz WH, Bowman BA, Marks JS, Koplan JP. The spread of the obesity epidemic in the United States, 1991-1998. *JAMA*. 1999;282:1519-1522.

Must A, Spadano J, Coakley EH, Field AE, Colditz G, Dietz WH. The disease burden associated with overweight and obesity. *JAMA*. 1999;282:1523-1529.

Stamler J. Epidemic obesity in the United States. *Arch Intern Med*. 1993;153:1040-1044.

Stevens J, Cai J, Pamuk ER, Williamson DF, Thun MJ, Wood JL. The effect of age on the association between body-mass index and mortality. *N Engl J Med*. 1998;338:1-7.

Taira K, Hikita M, Kobayashi J, et al. Delayed post-prandial lipid metabolism in subjects with intra-abdominal visceral fat accumulation. *Eur J Clin Invest*. 1999;29:301-308.

Wilson PW, Kannel WB, Silbershatz H, D'Agostino RB. Clustering of metabolic factors and coronary heart disease. *Arch Intern Med*. 1999;159:1104-1109.

Women

Barrett-Connor E, Bush TL. Estrogen and coronary heart disease in women. *JAMA*. 1991;265:1861-1867.

Bass KM, Newschaffer CJ, Klag MJ, Bush TL. Plasma lipoprotein levels as predictors of cardiovascular death in women. *Arch Intern Med.* 1993;153:2209-2216.

Cheng GS. Cardiac events increased in first two years of HRT. *Intern Med News.* 2000;33:1-2.

Delmas PD, Bjarnason NH, Mitlak BH, et al. Effects of raloxifene on bone mineral density, serum cholesterol concentrations, and uterine endometrium in postmenopausal women. *N Engl J Med.* 1997;337:1641-1647.

Godsland IF, Crook D, Simpson R, et al. The effects of different formulations of oral contraceptive agents on lipid and carbohydrate metabolism. *N Engl J Med.* 1990;323:1375-1381.

Goldman L, Goldman PA, Williams LW, Weinstein MC. Cost-effectiveness considerations in the treatment of heterozygous familial hypercholesterolemia with medications. *Am J Cardiol.* 1993;72:75D-79D.

Grady D, Rubin SM, Petitti DB, et al. Hormone therapy to prevent disease and prolong life in postmenopausal women. *Ann Intern Med.* 1992;117:1016-1037.

Grady D, Wenger NK, Herrington D, et al. Postmenopausal hormone therapy increases risk for venous thromboembolic disease. The Heart and Estrogen/Progestin Replacement Study. *Ann Intern Med.* 2000;132:689-696.

Granfone A, Campos H, McNamara JR, et al. Effects of estrogen replacement on plasma lipoproteins and apolipoproteins in postmenopausal, dyslipidemic women. *Metabolism.* 1992;41:1193-1198.

Hulley S, Grady D, Bush T, et al. Randomized trial of estrogen plus progestin for secondary prevention of coronary heart disease in postmenopausal women. Heart and Estrogen/progestin Replacement Study (HERS) Research Group. *JAMA.* 1998;280:605-613.

Kuhn FE, Rackley CE. Coronary artery disease in women. Risk factors, evaluation, treatment, and prevention. *Arch Intern Med.* 1993;153:2626-2636.

Manson JE, Willett WC, Stampfer MJ, et al. Body weight and mortality among women. *N Engl J Med.* 1995;333:677-685.

Nabulsi AA, Folsom AR, White A, et al. Association of hormone-replacement therapy with various cardiovascular risk factors in postmenopausal women. The Atherosclerosis Risk in Communities Study Investigators. *N Engl J Med.* 1993;328:1069-1075.

Sacks FM, Pfeffer MA, Moye LA, et al. The effect of pravastatin on coronary events after myocardial infarction in patients with average cholesterol levels. Cholesterol and Recurrent Events Trial investigators. *N Engl J Med.* 1996;335:1001-1009.

Stone NJ. Estrogen-induced pancreatitis: a caveat worth remembering. *J Lab Clin Med.* 1994;123:18-19.

Shlipak MG, Simon JA, Vittinghoff E, et al. Estrogen and progestin, lipoprotein(a), and the risk of recurrent coronary heart disease events after menopause. *JAMA.* 2000;283:1845-1852.

Taskinen MR, Puolakka J, Pyorala T, et al. Hormone replacement therapy lowers plasma Lp(a) concentrations. Comparison of cyclic transdermal and continuous estrogen-progestin regimens. *Arterioscler Thromb Vasc Biol.* 1996;16:1215-1221.

Thomas JL, Brans PA. Coronary artery disease in women. A historical perspective. *Arch Intern Med.* 1998;158:333-337.

Walsh BW, Sacks FM. Effects of low dose oral contraceptives on very low density and low density lipoprotein metabolism. *J Clin Invest.* 1993;91:2126-2132.

12 Treatment Algorithm

In the preceding chapters, the relevant information needed to manage patients with dyslipidemia has been discussed. This chapter attempts to outline a useful, overall approach to aid in the management of patients with lipid disorders.

The dyslipidemia should be categorized to determine if the goal of treatment is to lower the risk of coronary heart disease (CHD) and/or pancreatitis. Consider the following questions and approach for the patient who has been asked to obtain a fasting lipid profile (cholesterol, triglycerides, and high-density lipoprotein cholesterol [HDL-c], with determination of the low-density lipoprotein cholesterol [LDL-c] by formula). Evaluation of the degree of triglyceride excess, if present, helps develop an effective way to consider options for both workup and therapy.

1. **Triglycerides are under 200 mg/dL (or under 150 mg/dL in patients with diabetes and/or CHD with atherosclerosis)**.
   - Are the numbers believable? Is the patient fasting?
   - Determine CHD and risk-factor status to determine the risk of near-term CHD events to aid in setting LDL-c goals (Table 12.1).
   - LDL-c and HDL-c will be important to follow to determine success of therapy.
   - Proceed with workup of significant dyslipidemia:
     - Consider genetic disorders involving cholesterol-rich lipoproteins (Chapter 2).
     - Screen family for lipid disorders; problem may be expressed in childhood if LDL-c is

TABLE 12.1 — CORONARY HEART DISEASE RISK-FACTOR STATUS IN SETTING LDL-C GOALS
• Very high risk—those with coronary artery disease or peripheral vascular disease or atherosclerotic cerebrovascular disease – Low-density lipoprotein cholesterol (LDL-c) goal is <100 mg/dL • High risk—two or more risk factors, no overt coronary heart disease (CHD) – LDL-c goal is <130 mg/dL – Risk reduction is high priority; lipid-lowering drugs must be considered carefully in those at highest risk for CHD • Low risk—without two or more risk factors or CHD – LDL-c goal is <160 mg/dL – Avoid drug therapy to lower LDL-c if a patient's CHD risk is low

high enough to make familial hypercholesterolemia a likely possibility.
  - Careful examination, especially look for xanthomas or signs of secondary dyslipidemia (Chapters 3, 4).
  - Rule out secondary causes (Chapter 4).
- Nonpharmacologic treatment (Chapters 8, 9):
  - Emphasize a low saturated fat/dietary cholesterol regimen and avoiding calorie excess.
  - Prescribe regular aerobic exercise to raise HDL-c and keep triglycerides down.
- Pharmacologic treatment (Chapter 10):
  - Statins are the drugs of choice for lowering raised LDL-c despite diet.
  - For those with confirmed symptomatic CHD, statins may be started to achieve the more stringent LDL-c goal of 100 mg/dL or less. Resins are to be considered in young patients

with either familial hypercholesterolemia and/or a family history of premature CHD.
  - For those with the highest levels of LDL-c, combination therapy with statin and resin or even statin, resin, and niacin should be considered.
  - For certain refractory cases of heterozygous or homozygous familial hypercholesterolemia, LDL apheresis, partial ileal bypass, and liver transplantation are therapeutic options.

2. **Triglycerides are mildly to moderately elevated (200 to 400 mg/dL) or over 150 mg/dL in patients with diabetes and/or CHD with atherosclerosis.**

- Are the numbers believable? Is the patient fasting?
- Determine CHD and risk-factor status to determine the risk of near-term CHD events to aid in setting LDL-c goals (Table 12.1).
- Follow LDL-c; in this group, triglycerides and HDL-c must be followed closely as well. Note that direct measurement of LDL-c, cholesterol/HDL ratio, and/or non–HDL-c may be of value for assessing CHD risk when LDL-c is hard to determine by calculation due to triglyceride excess.
- Proceed with workup of significant dyslipidemia:
  - Consider genetic disorder involving both cholesterol and triglyceride-rich lipoproteins (Chapter 2).
  - Screen family for lipid disorders; problem may be expressed only in adult family members.
  - Perform careful examination, especially looking for xanthomas or signs of secondary dyslipidemia (Chapters 3, 4).
  - Rule out secondary causes (Chapter 4).

- Nonpharmacologic treatment (Chapters 8, 9):
  - Emphasize weight control, alcohol restriction, avoidance of sugars, increased complex carbohydrates, as well as low–saturated-fat/dietary-cholesterol regimen.
  - Prescribe regular aerobic exercise.
- Pharmacologic treatment (Chapter 10):
  - Niacin or statin therapy should be used. Some patients will require combination therapy (niacin and resin have the longest history of use). Niacin and statin are also a useful combination. The patient must be carefully monitored for liver toxicity or increased risk of myositis. Gemfibrozil and a statin are useful in those with combined hyperlipidemia and a very high CHD risk if other attempts to reach lipid goals fail. This combination requires careful follow-up with excellent patient compliance as there is increased risk of myositis and/or liver toxicity.

3. **Triglycerides are between 400 and 1000 mg/dL. Evaluate for risk of CHD and/or acute pancreatitis (must avoid exacerbating hypertriglyceridemia with diet and/or drugs).**
   - Are the numbers believable? Is the patient fasting?
   - Determine CHD and risk-factor status to determine the risk of near-term CHD events to aid in setting LDL-c goals (Table 12.1).
   - It is difficult to determine CHD risk when triglycerides are between 400 and 1000 mg/dL.
     - LDL-c cannot be determined by formula due to elevated triglycerides.
       - Could measure LDL-c directly or
       - Could use non–HDL-c (cholesterol–HDL-c) to aid in coronary-risk assessment.

- Attempt to lower triglycerides with non-pharmacologic means and then measure LDL-c by formula (not unreasonable since triglycerides are too high).

- Proceed with workup of patient with significant dyslipidemia:
  - Consider genetic disorder involving triglyceride-rich lipoproteins (Chapter 2).
  - Screen family for lipid disorders; problem may be expressed only in adult family members.
  - Do a careful examination, especially looking for xanthomas or signs of secondary dyslipidemia (Chapters 3, 4).
  - Rule out secondary causes (Chapter 4).
- Nonpharmacologic treatment (Chapters 8, 9):
  - Emphasize weight control, alcohol restriction, avoidance of sugars, increased complex carbohydrates, as well as low–saturated-fat/dietary-cholesterol regimen.
  - Prescribe regular aerobic exercise.
- Pharmacologic treatment (Chapter 10)
  - Gemfibrozil is the drug of choice if nonpharmacologic therapy alone cannot reduce fasting triglyceride level to <700 mg/dL.
  - For those with triglycerides under 500 mg/dL, niacin or a statin might prove useful.

4. **Is the triglyceride level more than 1000 mg/dL? In these patients, the risk is acute pancreatitis.**

- Are the numbers believable? Was the patient fasting or nonfasting?
  - Even if nonfasting, triglycerides are greatly in excess.
  - Put a tube of the patient's plasma in the refrigerator and see if a creamy supernatant forms to confirm fasting chylomicronemia (Chapter 3).

  - If the patient has abdominal pain, associated pancreatitis must be ruled out.
- Rule out secondary causes (Chapter 4):
  - Diet: high fat, high calories, alcohol excess
  - Drugs: estrogen, steroids
  - Disorders of metabolism: low thyroid, diabetes, pregnancy
  - Diseases: nephrosis, systemic lupus erythematosus
- Consider genetic disease (Chapters 2, 3)
- Clinical examination (Chapter 3):
  - Eruptive xanthomas, lipemia retinalis, hepatosplenomegaly will occur in some patients.
  - Suspect pancreatitis even if amylase is not elevated.
- Nonpharmacologic therapy (Chapters 8, 9):
  - Low-fat diet is crucial part of therapy.
    - Nothing by mouth (NPO) if acute pancreatitis.
    - Very low-fat diet should be adhered to until triglycerides are clearly falling below 1500 mg/dL.
    - Dietary fat should be cautiously increased based on triglyceride level; may consider fish oil and medium-chain triglyceride oil.
  - Regular aerobic exercise, once the patient is active again, will help control weight and improve triglyceride concentration.
- Pharmacologic therapy (Chapter 10):
  - Gemfibrozil is the drug of choice in patients with marked triglyceride excess.
  - Statins or resins are not appropriate in patients who present with triglyceride values at this high level.

5. **Isolated low HDL-c.**
   - Are the numbers believable? Is the patient fasting?

- Must carefully assess risk of CHD as not all patients with low HDL-c are at high risk.
  - Vegetarians with low LDL-c/low blood pressure have low overall risk of CHD despite low HDL.
  - Some populations with isolated low HDL-c have higher risks of CHD.
    - A particularly high-risk group is one with a family history of premature CHD and low HDL-c.
    - HDL-c is a strong predictor of CHD status in those over age 50.
- See Table 12.1 to determine CHD risk.
- For those at risk for CHD:
  - Consider genetic syndromes and screen family (Chapters 2, 3).
  - Rule out secondary causes of low HDL-c (Chapter 4).
  - Nonpharmacologic approach (Chapters 8, 9):
    - Lose excess weight.
    - Stop cigarette smoking.
    - Avoid simple sugars.
    - Consider lower carbohydrate, higher monounsaturated fat intake, especially in the diabetic patient.
    - Exercise regularly.
  - Pharmacologic approach (Chapter 10):
    - Niacin is the drug of choice for raising HDL-c.
    - HDL-c often is refractory; consider lowering LDL-c with a statin.
    - Gemfibrozil therapy is problematic; it may raise HDL-c but also can raise LDL-c.

Although aggressive treatment of LDL-c is a high priority in secondary prevention of CHD and stroke, there is still debate as to the wisdom of using drugs to lower LDL in primary prevention. The following is an approach designed to "tease out" the high-risk patients, using available data.

Step I: Assess the baseline risk of an atherosclerotic event. Consider:

1. Fasting lipid profile: Is it high risk?
2. Risk-factor status: Is the patient at high risk for a near-term event?

a) One approach is to use comparable clinical trial data. For example, would the patient have qualified for the Air Force/Texas Coronary Atherosclerosis Prevention Study (AFCAPS/TexCAPS)? This would mean age 45 and older for men and 55 and older for women, LDL-c values in the range of 130 to 190 mg/dL and HDL-c values $\leq$45 for men and $\leq$47 for women. Treating all of those who would qualify would greatly expand the number receiving statins. If cost-benefit is considered, some may not find this approach to be cost-effective due to the lower rate of CHD events in the placebo group in this trial.

b) Another approach is to look at categorical risk factors. For example, multiple risk factors related to insulin resistance would increase risk greatly, particularly after age 50 in men and 60 in women. A striking family history of premature CHD might be another factor that would influence clinical judgment. Lp(a) in a postmenopausal woman might influence hormone replacement therapy (see Chapter 11, *Special Populations*).

c) A third approach is a quantitative approach using a risk-scoring technique based on the summation of graded risk factors as advocated by Framingham investigators (Wilson et al, 1998). The Writing Group for Prevention V felt that strong consideration should be given to such an approach, although there was concern about its practicality. The physician or his/her assistant can easily score a sheet of paper with the Framingham algorithm. One caution is the use of this tool in certain populations that are nonwhite or non-European. Also it may underestimate the effect of severe expressions of risk factors (eg, familial hypercholesterolemia, heavy cigarette smoking).

3. Subclinical assessment of disease. The Writing Group for Prevention V noted that at every level of risk factor exposure, there is substantial variation in the amount of atherosclerosis. This clinical variability is multifactorial. Genetic susceptibility, combinations and interactions with other risk factors, and the level and duration of exposure to each risk factor play a role. Also biologic and laboratory variability must be recognized. Measures of subclinical CHD may enhance CHD-risk prediction in individual cases, but barriers to their routine use remain. These include cost-effectiveness, adequate clinical trial data, availability of the test, and standardization of the methods used. Currently, noninvasive tests such as carotid artery duplex scanning, electron-beam computed tomography, ankle/brachial blood pressure ratios, and high-sensitivity C-reactive protein (hs-CRP) are available to clinicians. The ankle- brachial index (ABI) <0.90 in either leg is considered evidence of peripheral arterial disease, and progressively lower ABI values indicate more severe obstruction. This correlates strongly with coronary atherosclerosis in individuals over 50. Carotid artery duplex scanning is noninvasive and can add incremental information, but it requires an experienced laboratory. Coronary cal-

cium scores in asymptomatic subjects should be interpreted carefully. Although coronary calcium correlates with atherosclerosis, the concern is that the good clinical studies are still not at hand. Prevention V felt that its best use was in detecting extensive atherosclerosis in those at intermediate risk as this might change therapy. Thus its selective use in these patients may be of value. Certainly, an older diabetic with risk factors who is likely to have extensive calcium is unlikely to benefit, as is a young man or woman without risk factors. Emerging data suggest value of the hs-CRP (not the usual test performed by clinical labs, but a new, more sensitive test) to predict CHD in middle-aged men and women. The test is standardized and its prediction is over and above that determined by the cholesterol/HDL-c ratio (Ridker et al, 1998).

4. Exercise testing in hypercholesterolemic men can help stratify as well. In the Lipid Research Clinics Mortality Follow-up study, the results of a positive exercise treadmill test (1 mm or more of ST depression) in 3600 men 35 to 79 years of age without prior myocardial infarction (MI) were striking. Cumulative mortality from cardiovascular disease was 11.9% (22/185) in the 8.1 years of mean follow-up among men with a positive exercise test as contrasted with 1.2% (36/2993) among men with a negative test. The data suggested that a positive test was equivalent to a 17.4 year increment in age.
5. Safety considerations. The long-term nature of cholesterol-lowering drug therapy makes safety considerations paramount. Thus nonpharmacologic therapy (Chapters 8 and 9) should be utilized in both low-risk and high-risk individuals. In the former case, diet and exercise might

reduce the risk-factor profile and forestall disease. In the latter case, there is the promise of reducing the dose of medication as well as treating associated abnormalities in HDL-c and triglycerides.

Step II: If the patient is at deemed high enough risk to treat, consider the following: (McQuay and Moore, 1997)

1. Could you easily treat the event you are trying to prevent if it occurred? Answer: Sudden death can be the first event of CHD in susceptible people in the community.
2. Would the event you are trying to treat be a serious event? Answer: Yes—sudden death, MI, and unstable angina are serious events.
3. Does the prophylaxis have adverse effects? Answer: Statin drugs seem safe over the intermediate term (see Chapter 7).
4. Is the prophylaxis effective (eg, have a low number-needed-to-treat score)? Answer: See Table 7.12.

SUGGESTED READINGS

12

Gordon DJ, Ekelund LG, Karon JM, et al. Predictive value of the exercise tolerance test for mortality in North American men: the Lipid Research Clinics Mortality Follow-up Study. *Circulation.* 1986;74:252-261.

Greenland P, Abrams J, Aurigemma GP, et al. Prevention Conference V: Beyond secondary prevention: identifying the high-risk patient for primary prevention: noninvasive tests of atherosclerotic burden: Writing Group III. *Circulation.* 2000;101:E16-E22.

Grundy SM, Bazzarre T, Cleeman J, et al. Prevention Conference V: Beyond secondary prevention: identifying the high-risk patient for primary prevention: medical office assessment: Writing Group I. *Circulation.* 2000;101:E3-E11.

McQuay HJ, Moore RA. Using numerical results from systematic reviews in clinical practice. *Ann Intern Med*. 1997;126:712-720.

Ridker PM, Glynn RJ, Hennekens CH. C-reactive protein adds to the predictive value of total and HDL cholesterol in determining risk of first myocardial infarction. *Circulation*. 1998;97:2007-2011.

Schachinger V, Halle M, Minners J, Berg A, Zeiher AM. Lipoprotein(a) selectively impairs receptor-mediated endothelial vasodilator function of the human coronary circulation. *J Am Coll Cardiol*. 1997;30:927-934.

Wilson PW, D'Agostino RB, Levy D, Belanger AM, Silbershatz H, Kannel WB. Prediction of coronary heart disease using risk factor categories. *Circulation*. 1998;97:1837-1847.

INDEX

Note: Page numbers in *italics* indicate figures; page numbers followed by t indicate tables.

13

13

13

13

13

13